Content

Page No.

PART-1

INTRODUCTION

What is a Project?

It's estimated that half of our work life is spent on routine, repetitive tasks: processing time cards, filling our sales orders, picking up passengers, and delivering parcels. Projects account for the other half of the work done in organizations. A project is a job that has a beginning and an end (time), a specified outcome (scope) at a stated level of quality (performance), and a budget (costs). Hence are a few examples of projects:

- Moving your company's offices to new location
- Developing a new software application
- Creating a policy manual
- Remodelling a room in your home
- Developing an internet
- Preparing for accreditation or certification
- Revamping a training program
- Starting a new business
- Auditing your organization's software or accounting systems

The four project parameters- time, scope, performance, and costs- are related. This relationship is often expressed as a formula [C=f(P,T,S)], which indicates that a project's cost is a function of a project's time, scope, and performance. One definition of project management is efficiently using resources to complete a project as designed, on time, at the desired level of performance, and within budget. These project parameters are also called constraints.

- Performance: "We have an additional crew that hasn't built a deck before, but they've got some extra time on their hands, so we can assign them to help".
- Scope: "We can't build the benches in time, and the deck won't be stained or treated".
- Costs: "We can build the deck as planned, but it cost extra.

PART-2

PROJECT

Creating a Task List

In this chapter, you will learn how to:

- ✓ Start Microsoft Project standard or professional and save a new project plan.
- ✓ Enter task names.
- ✓ Estimate and record how long each task should last.
- ✓ Create a milestone to track an important event.
- ✓ Organize task into phases.
- ✓ Create task relationships by linking tasks.
- ✓ Switch task scheduling from manual to automatic.
- ✓ Set non-working days for the project plan.
- ✓ Check the project plan's overall duration.
- ✓ Record task details in notes and insert a hyperlink to content on the web.

Tasks are the most basic building blocks of any project—tasks represent the work to be done to accomplish the goals of the project. Tasks describe project work in terms of sequence, duration, and resource requirements. In project, there are several different kinds of tasks. These include summery tasks, subtasks and milestone (all discussed in this chapter). More broadly, what are called tasks in Project [1]is sometimes also called activities or work packages.

In this chapter, you will manage the scheduling of tasks in two different ways:

- Enter tasks as manually scheduled to quickly capture some details without actually scheduling tasks.
- Work with automatically scheduled tasks to begin to take advantage of the powerful scheduling engine in project.

Creating a new project plan

A project plan is essentially a model that you construct of some aspects of a project you are anticipating – what you think will happen, or what you want to happen (it's usually best if these are not too different). This model focuses on some, but not all, aspects of the real projects- tasks, resources, time frame, and possibly their associated costs.

As you might expect, Project focuses primarily on time. Sometimes you might know the planned start date of a project, the planned finish date, or both. However, when working with project, you specify only one date, not both: the

[1] How to get MS Project?

Download MS Project from following link

http://www.microsoft.com/office/project/

project start date or the project finish date. Why? Because after you enter the project start or finish date and the durations of the tasks, Project calculates the other date for you.

Remember that Project is not just merely a static repository of your schedule information or a Gantt chart drawing tool; it is an active scheduling engine.

Most projects should be scheduled from a start date, even if you know that the project should finish by a certain deadline date. Scheduling from a start causes all tasks to start as soon as possible, and it gives you the greatest scheduling flexibility. In this, you will see flexibility in action as we work with a project that is scheduled from a start date.

Creating a new project plan, set its start date, and save it.

SET UP Start Project if it's not already running.

1. Click the File tab.

   Project displays the backstage view.

2. Click the new tab.

   Project displays your options for creating a new project plan.

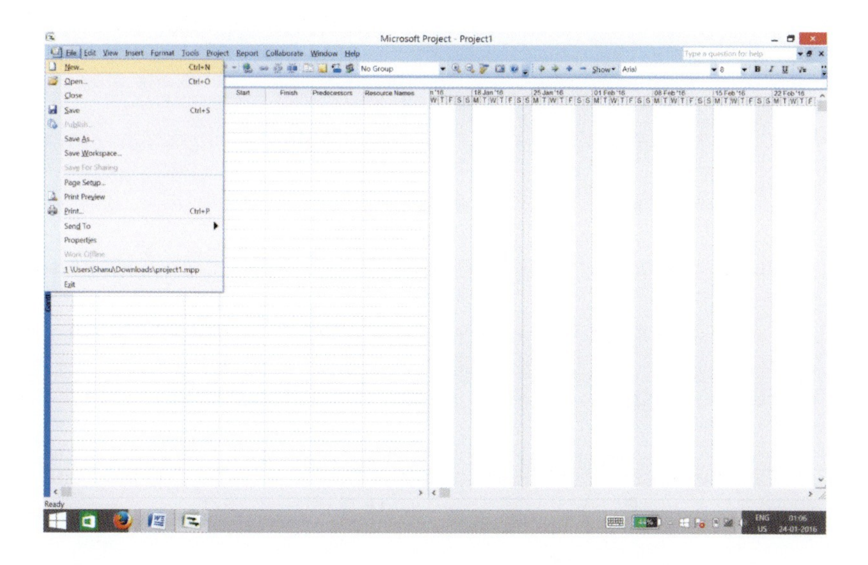

These options include using templates installed with Project or available on the web. For this exercise, you will create a new blank project plan.

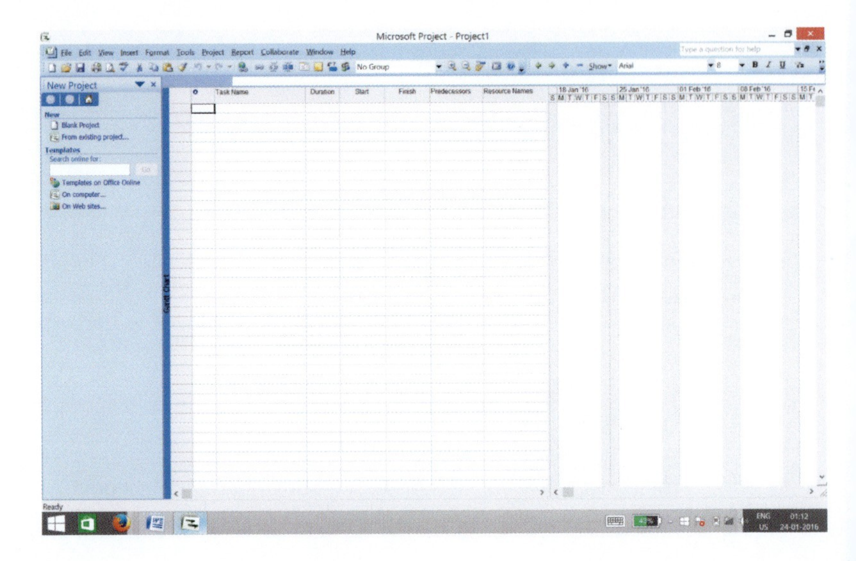

3. Under available templates, make sure that blank project is selected, and then click the Create button on the right side of the backstage view.

   Project creates a new project plan. You may see a note briefly reminding you that new tasks are created in manually scheduled mode. This information remains visible on the status bar as well.

4. On the Project tab, in the properties group, click Project Information.

   The Project Information dialog box appears.

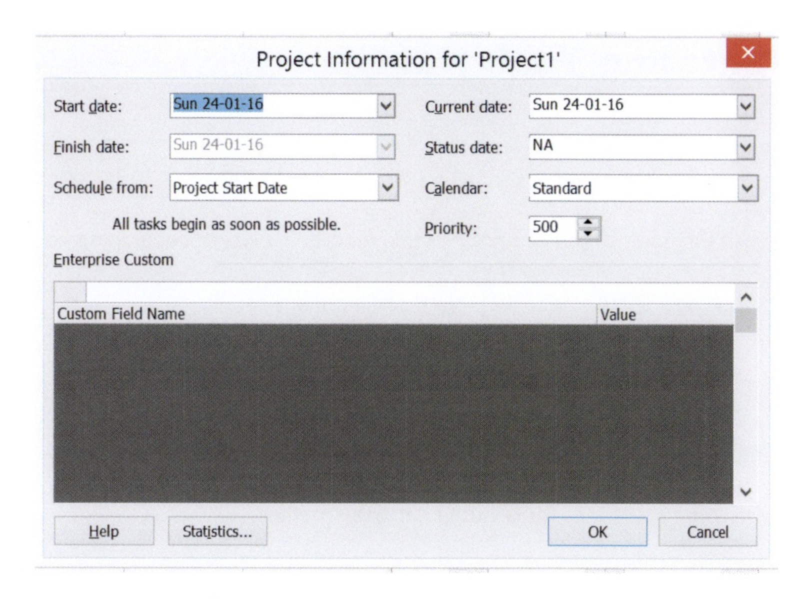

5. In the start date box, type 24-01-16, or click the down arrow to display the calendar and select January 24, 2016.

   Tip In the calendar, you can navigate to any month and then click the date you want, or click today to quickly choose the current date.

6. Click OK to accept this start date and close the Project Information dialog box.

7. On the File tab, click Save.

   Because this project plan has not been previously saved, the Save As dialog box appears.

Entering Task Names

As mentioned previously, tasks represent the work to be done to accomplish the goals of the project. For this reason, it's worth developing good practices about how you name tasks in your project plans.

Task names should be recognizable and make sense to the people who will perform the tasks and to other shareholders who will see the task names. Here are some guidelines for creating good task names.

- Use short verb phases that describe the work to be done, such as "Edit manuscript".
- If tasks will be organised into phases, don't repeat details from the summary task name in the subtask name unless it adds clarity.
- If tasks will have resources assigned to them, don't include resource names in the task names.

Keep in mind that you can always edit task names later, so don't worry about getting exactly the right task names when you're initially entering them into a project plan. Do aim to use concise, descriptive phrases that communicate the required work and make sense to you and others who will perform the work.

PART-3

SCHEDULING OF PROJECT

How scheduling works in Project

Microsoft office project uses a powerful scheduling engine to help align your organizations projects and tasks with available resources. Understanding how Project schedules work is a key factor in making intelligent planning decisions.

➢ **How is a Project scheduled?**

Project schedules a project from the information that you enter about the following:

- The overall project.
- The individual work items (called tasks) required to complete the project.
- If necessary, the resources needed to complete those tasks.

If anything about your project changes after you creates your schedule, you can update the tasks or resources and Project adjusts the schedule for you.

For each task, you might enter one or all of the following:

- Durations
- Task dependencies
- Constraints

Using this information, Project calculates the start date and finish date for each task.

You can enter resources in your project and then assign them to tasks to indicate which resource is responsible for completing each assignment. Not only does this help you plan project staffing, it can also help you to calculate the number of machines needed or the quantity of material to be consumed. If you enter resources, task schedules are further refined according to the following resources information:

- Work
- Units
- working times entered in calendars

Other elements, such as lead time and lag time, task types, resource availability, and the driving resource, can affect scheduling, so understanding the effects of these elements can help you to maintain and adjust your schedule as needed.

➢ How does the project start date affect the schedule?

If you enter a start date for the project, by default, Project schedules tasks to begin on the project's start date and calculates the project's finish date based on the last task to finish. As you enter more information about tasks, such as task dependencies, durations, and constraints, Project adjusts the schedule to reflect more accurate dates for the tasks.

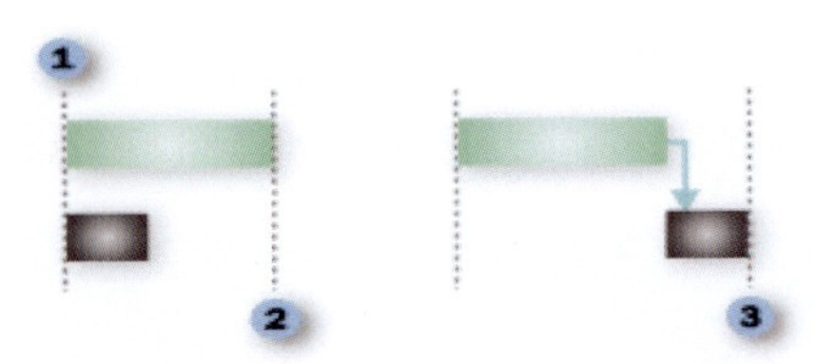

1. When you create a new project, you first enter the project's start date. When you schedule a project from the start date, all tasks start at the project start date unless you specify otherwise.

2. With no task dependencies or constraints applied, the project's duration is the same as the duration of the longest task. In other words, the project finish date is the same as the longest task's finish date.

3. Task dependencies, such as the finish-to-start dependency between the first and second tasks (as shown here), can change the project's finish date.

Nearly all projects should be scheduled from a known start date. Even if you know the date that a project must be completed, scheduling from a start date gives you the maximum flexibility.

However, you might want to schedule from a finish date when:

- You need to determine when a project must start so that it finishes on a specific required date.
- You are not sure when your project will begin (for example, you are receiving work from another source that could be delayed).
- You project management methodology requires you to schedule from a finish date.

As you work with your project that is scheduled from a finish date, be aware of differences in the way that Project handles some actions:

- When you enter a task, Project automatically assign the As Late Possible As (ALAP) constraint to the finish date of the task. You should set other constraints only when necessary.
- If you drag a Gantt bar to change finish date of a task, Project automatically assign a Finish No Later Than (FNLT) constraint.

- If you change your project to schedule from a finish date and it was previously scheduled from a start date, you will remove all leveling delays and leveling splits from tasks and assignments.
- If you use automatic leveling to reduce resource over allocations in your projects, Project will add a leveling delay after a task rather than before a task.

- **What are the default settings for calculating the schedule?**

Project calculates the duration of tasks based on the definitions of the duration units in the **Calendar** tab of the **Options** dialog box (Tools menu). Just like a normal monthly calendar, the year begins in January and each week begins on Sunday or Monday. By default, when Project calculates duration units, one day equals 8 hours, one week equals 40 hours, and one month equals 20 working days. If you enter and finish dated for tasks and don't enter start and finish times, Project uses 8:00 A.M. as the default start time and 5:00 P.M. as the default and time.

- **How do constraints affect the schedule?**

When you need to control the start or finish date of a task, you can change the constraint on the task. Flexible constraints work with task dependencies to make a task as soon or as late as the task dependencies will allow. For example, a task will an As Soon As Possible (ASSP) constraint and a finish-to-start dependency will be scheduled as soon as the predecessor task finishes. By default, all tasks in as project that is scheduled from the start date have the ASAP constraint applied. Likewise, by default, all tasks in a project that is scheduled from the finish date have the As Late As Possible (ALAP) constraint applied.

Constraints with moderate scheduling flexibility will restrict a task from starting or finish before or after a day you choose. For example, a task with a Start No Earlier Than (SNET) constraint for June 15 and finish-to-start dependency to another task can begin June 15 if its predecessor is finished by June 15 (or later if its predecessor finishes after June15), but it can't be scheduled before June 15.

1. With the default finish-to start task relationship and as ASAP constraint applied to these tasks, the successor task (the second one) is scheduled to begin as soon as the predecessor task (the first one) is schedules to finish.
2. With a SNET constraint applied, the successor task cannot begin before the constraint date, even if (As shown here) the predecessor task is completed before the constraint date.

Inflexible constraints override any task dependencies by default and restrict a task to a date you choose. For example, a task with a Must Start On (MSO) constraint for September 30 and a finish-to-start dependency to another task will always be scheduled for September 30 no matter whether its predecessor finishes early or late.

If a task that is constrained to a date has a predecessor that finishes too late for the successor to begin on the date specified in the constraint, negative slack can occur.

- **How do deadline dates affect the schedule?**

Deadline dates don't usually affect task scheduling. They are used to indicate a target date you don't want to miss, without requiring you to set a task constraint that could affect scheduling if predecessor tasks change. A task with a deadline is scheduled just like any other task, but when a task finishes after its deadline, Project displays a task indicator notifying you that the task missed its deadline.

Deadline dates can affect the total slack on tasks. If you enter a deadline date before the end of the task's total slack, total slack will be recalculated by using the deadline date rather than the task's late finish date. The task becomes critical if the total slack reaches zero.

You can set deadlines for summary tasks as well as individual tasks. If the summary task's deadline conflicts with any of the subtasks, the deadline indicator signifies a missed deadline among the subtasks.

But deadline dates can affect how tasks are scheduled if you set a deadline date on a task with an As Late As Possible (ALAP) constraint. The task is scheduled to finish on the deadline date, though the task could still finish after its deadline if its predecessors slipped.

- **How do calendars affect the schedule?**

Calendars determine the standard working time and nonworking time, such as weekends and holidays, for the project. They are used to determine the resource availability, how resources that are assigned to tasks are scheduled, and how the tasks themselves are scheduled. Project and task calendars are used in scheduling the tasks, and if resources are assigned to tasks, resource calendars are used as well.

The calendars referred to in Project are:

- **Base calendars** These are the foundations for the other types of calendars. You can also choose a base calendar to be the project calendar, and you can apply a base calendar to tasks as a task calendar or as the default hours for a resource calendar. Project provides three base calendars: the Standard, 24-Hours, and Night Shift calendars. You can customize your own base calendar by using any of the base calendars provided.

- **Project calendars** These set the standard working and nonworking times for the project as a whole. If resource calendars or task calendars are not used, tasks are scheduled during the working time on the project calendar by default.
- **Resource calendars** These are based on the Standard calendar by default. You can change working time or nonworking time for specific resources or a set of resources, ensuring that resources are scheduled only when they are available for work. If you have changed working or nonworking time on a resource calendar and the resource is assigned to a task, the task is scheduled during the working time on the resource calendar.
- **Task calendars** These can be used to define working times for tasks outside the working times in the project calendar. When a task calendar is assigned to a task and the resource assigned to the task has different working times in its resource calendar, the task is scheduled for the overlapping working time of the two calendars. But you can set a

task option to ignore resource calendars and schedule the task through the resource's nonworking time.

- **How do resource assignments drive the schedule?**

If you don't assign resources to tasks in your project, Project calculates the schedule using durations, task dependencies, constraints, and project and task calendar information. If you do assign resources, the tasks are also scheduled according to resources' calendars and assignment units, providing for more accurate scheduling.

An assignment is the association of a specific task with a specific resource that is responsible for completing the task. More than one resource can be assigned to a task. Work resources, material resources, and cost resources can be assigned to tasks. Unlike work resources, assigning material resources or cost resources to a task does not affect task scheduling.

For example, in your project you have a task named **Develop specifications**. You also have an engineering resource, Sean. If you assign Sean to the **Develop specifications** task, the scheduling of this task depends on Sean's resource calendar and assignment units, in addition to task information such as duration, task dependencies, constraints, and calendars.

In addition to scheduling according to task information, after you assign resources to the tasks in your project, Project has resource and assignment information to use in calculating schedule information, including:

- The amount of work or overtime work the resource is assigned to do, and how that work is distributed over time. Work distribution over time can also be affected by work contours.
- The number of assignment units for the resource, that is, part-time, full-time, or multiple, on the task.
- The task type, which affects how a schedule changes if you revise the existing assignment. The three task types are fixed unit, fixed duration, and fixed work.
- Whether the task is effort-driven. If a task is effort-driven, as resources are added or removed on the assignment, the work remains constant for the task and is redistributed among the resources. For fixed-unit tasks, for example, one result is that if more resources are assigned, a shorter duration is required to complete the task.
- Resource calendars. Project schedules the assigned resources based on the working and nonworking times indicated on their resource calendars.

- **What information can help me analyze my project's progress?**

Five pieces of task information help you analyze progress as you track tasks in your project: duration, work, start date, finish date, and cost.

Variations of each of these types of fields help you compare and evaluate your progress: planned, scheduled, actual, and remaining.

For example, for one task, there can be fields of information containing planned work, scheduled work, actual work, and remaining work. The contents of these fields might match one another, or they might all be different. Variances between certain fields can also be examined for useful tracking information. For this reason, these fields are referred to as tracking fields.

PART-4

DEFINING THE PROJECT

Defining the Project

There are three activities that should complete before begin entering individual project tasks: setting file properties, entering project working times, and adding project properties. You can do these activities in any order, but they must be completes prior to other tasks. Although the first one, setting properties for the project file, is optional, you may already be in the habit of doing this when you start a file in other application. Choose File→ Properties from the menu option the Properties dialog box shown here. Make sure that you're listed as the Author. Enter a few keywords and a summary if you wish.

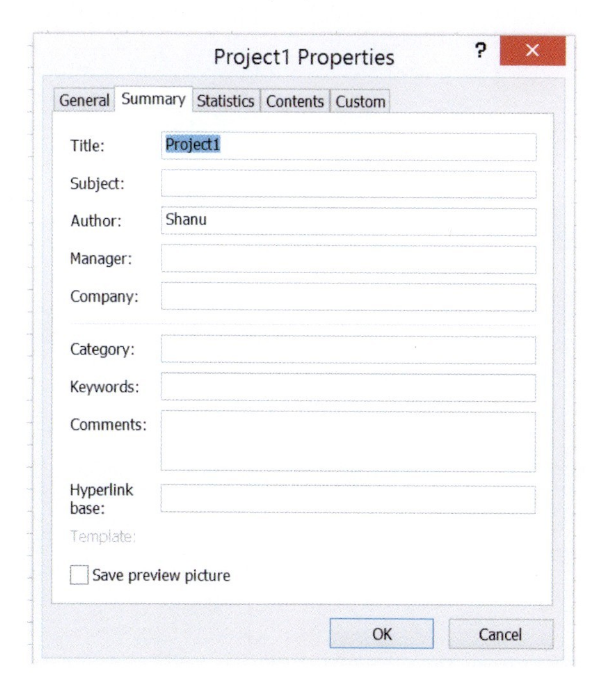

Entering the task list

Enter your project tasks roughly in order (although you can drag and drop tasks to rearrange them later) in MS Project's Gantt Chart view. To enter a task, either click in the Task Name textbox and enter the task's name, or double-click in the Task Name textbox to open a Task information dialog box. Enter the task name on the General tab.

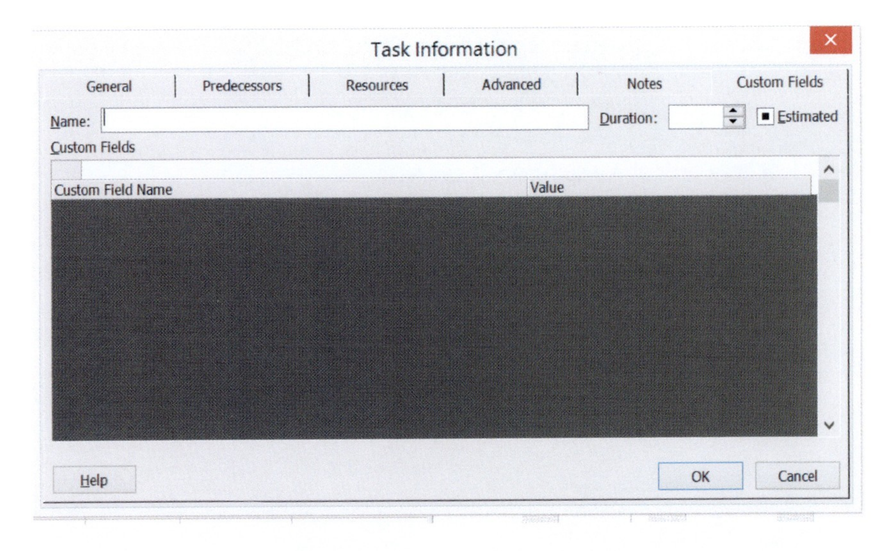

If you are entering tasks directly in Gantt Chart view, you can adjust the Gantt chart task columns, as you would with columns in Excel or Access, by pointing to the right edge of the column header button. The pointer changes to a column adjustment tool. Drag to the right to widen the column, or double- click to have Project size the column to fit the contents. Drag the vertical gray bar between the task columns and the Gantt Chart to display additional columns of task information.

PART-5

DEFINING THE PROJECT CALENDAR

Defining the Project Calender

Project uses several different types of calendars to reflect when resources are available to work on a project. All projects that you create are assigned a calendar by Project. The project calendar reflects the general working hours for the project. The calendar contains the number of hours per day that will be spent on the project and the days of the week when the work will occur. In the project calendar, you can exclude holidays and other days in which no work will be done, and set realistic expectations of how much time will be available from different resources, you can create additional calendars that can be assigned to specific tasks and particular resources.

Setting the standard calendar

The standard calendar is set for a 40-hour week: Monday through Friday, from 8:00 a.m. to 5:00 p.m., with one hour for lunch. No holidays are represented in the standard calendar. If this does not represent your standard schedule, you can change the settings for the standard calendar or any of the calendars or any of the calendars used in Project. To change the calendar settings, click Tools→ Change Working Time. This opens the Change Working Time dialog box.

The Change Working Time dialog box displays the calendar you chose in the Project Information dialog box. We selected the standard calendar, so it is the calendar that displays in the figure. Days that appear in the calendar in white are default working days.

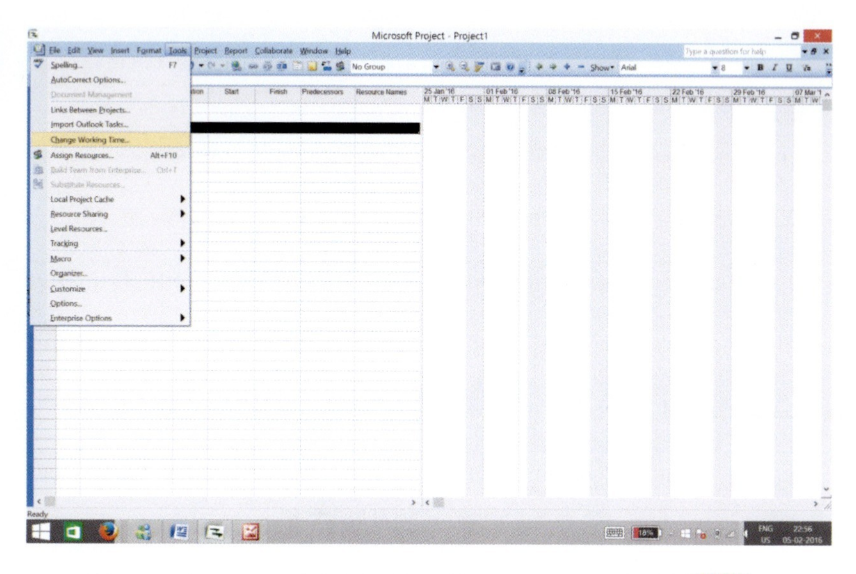
Microsoft Project - Project1
File Edit View Insert Format Tools Project Report Collaborate Window Help
Spelling... F7
AutoCorrect Options...
Links Between Projects...
Import Outlook Tasks...
Change Working Time...
Assign Resources... Alt+F10
Local Project Cache
Resource Sharing
Level Resources...
Tracking
Macro
Organizer...
Customize
Options...
Enterprise Options
Start
Finish
Predecessors
Resource Names
Ready

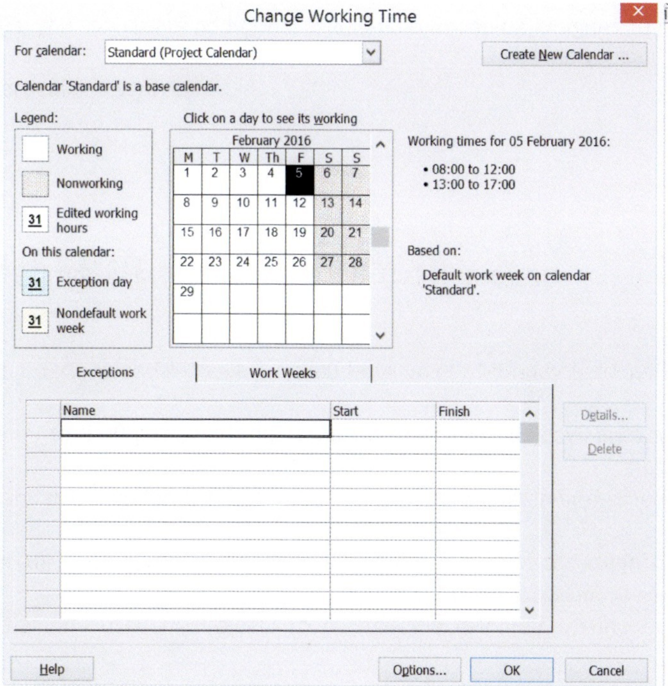
Change Working Time
For calendar:
Standard (Project Calendar)
Create New Calendar ...
Calendar 'Standard' is a base calendar.
Legend:
Working
Nonworking
Edited working hours
On this calendar:
Exception day
Nondefault work week
Click on a day to see its working
February 2016
Working times for 05 February 2016:
• 08:00 to 12:00
• 13:00 to 17:00
Based on:
Default work week on calendar 'Standard'.
Exceptions
Work Weeks
Name
Start
Finish
Details...
Delete
Help
Options...
OK
Cancel

PART-6

GANTT CHART

Gantt Chart

- Select Gantt Chart view from the view menu.

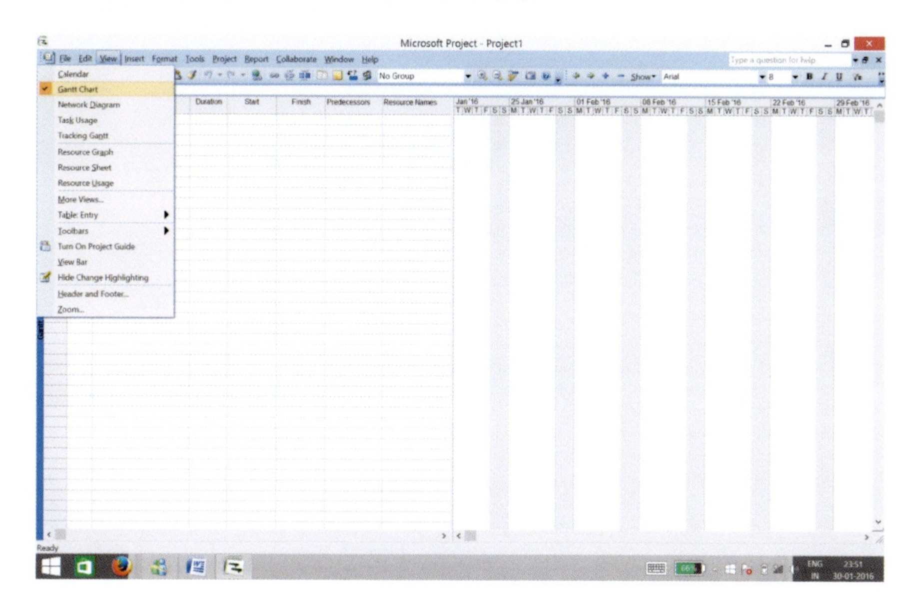

- You'll have a spreadsheet where you can now enter information of all the activities

  i.e. task name, duration, start date, end date, predecessors and various other fields.

- You can enter required information in two ways, in spreadsheet or when you double click on cell you get pop up window in which you can enter all the information of that particular activity.

- For predecessor activity you need to write activity number.

- The SW itself will calculate start and end date.
- Now the Gantt Chart is complete.

PART-7

TASKS

Tasks

A Project usually consists of four major types of tasks:

- Summary tasks are tasks that contain subtasks. Microsoft Project automatically summarizes the durations and costs related to subtasks into the summary task.
- Subtasks are smaller tasks that roll up into a summary task.
- Recurring tasks are tasks that occur at regular intervals during the course of the project- a project review meeting, for example.
- Milestones are tasks that usually have no duration and mark the completion of a significant phase of the project. Move completed might be an example of a milestone in the office space example used earlier.

When you create a task list, you should consider what type each task is, and be sure to include recurring tasks and milestones to make the list complete.

Adding tasks and milestones to a Project File

1. On the View menu, click Gantt Chart.
2. In the Task Name field, type a task name, and then press TAB. (Microsoft Project enters an estimated duration of one day for the task followed by a question mark)
3. In the Duration field, type the amount of time each task will take in months, weeks, days, hours, or minutes, not counting nonworking time. (By default the time period will be days, but that can be changed to hours, months, etc.)
4. Press ENTER.
5. It should look like the figure below:

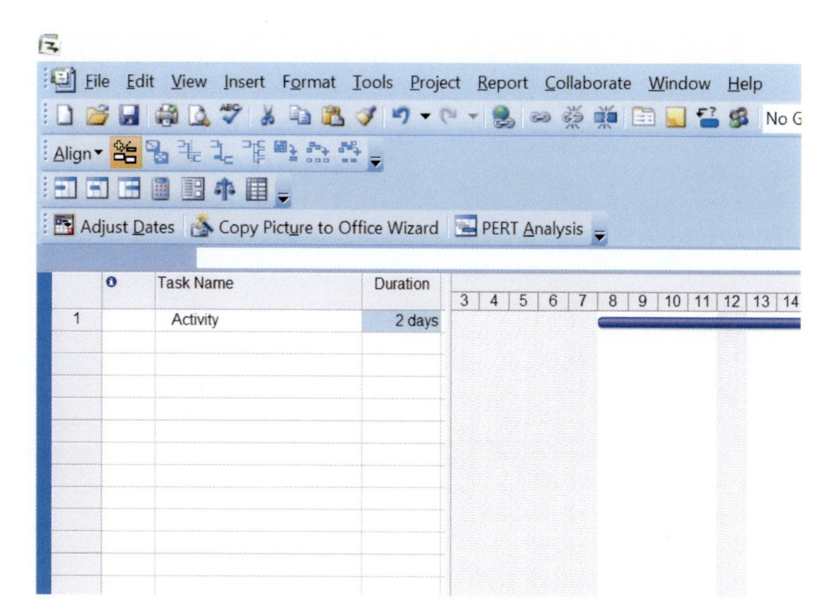

6. To add a milestone the only difference is that the duration of the activity must be zero (below is an example):

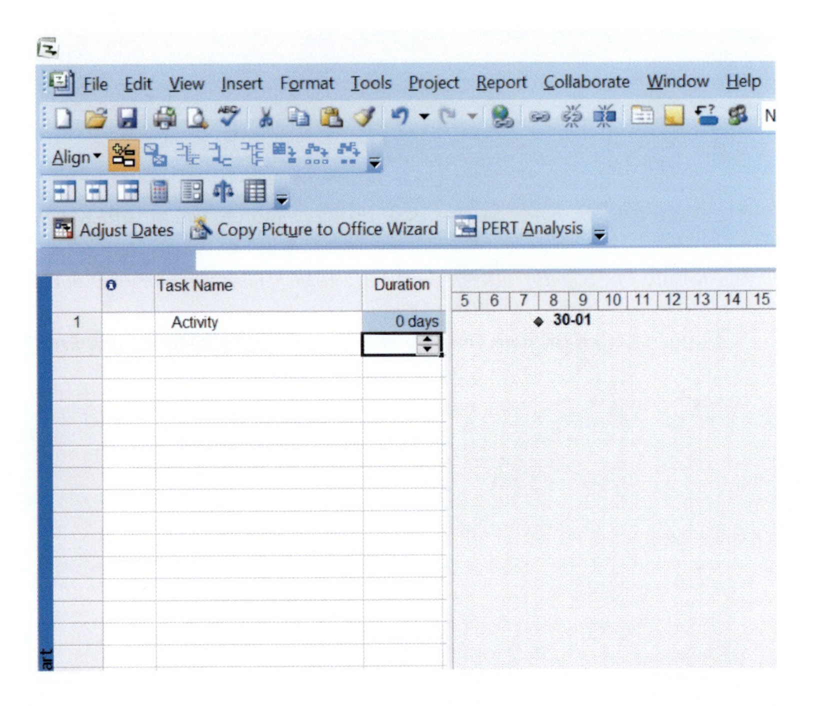

Note: By double clicking on a Task or milestone, you can modify its information with a form that prompts

Editing Tasks

After you enter a task list, chances are you will want to make some changes in it. You can edit text in a cell, adjust columns and rows, move and copy tasks, and insert and delete tasks.

Editing Text

Editing text in Gantt Chart view is more easily accomplished using the Entry bar. The Entry bar is the bar above the task list that displays entries you make to cells in the table. When you are entering or editing text, a Cancel button (X) and an Enter button (red check) appear in the Entry bar.

To edit in the Entry bar, follow these steps:

1. Click the cell you want to edit
2. Point to the contents in the Entry bar so the pointer changes to an I-beam.
3. Click the I-beam where you want to edit.
4. When you finish editing, click the Enter button (or press Enter) to accept the changes, or click the Cancel button to cancel the changes.

Undoing Mistakes

When you change your mind, Project gives you a chance to undo your last action. Click the undo button or choose Edit→ Undo.

When you click on Undo, then the button changes to a Redo button. If you want to reverse the last undo, click Redo or choose Edit→ Redo.

Moving and Coping Tasks

Being able to change the order of tasks is essential to effectively organizing a project. In Project, you can move and copy tasks by using the traditional cut/copy and paste, or drag and drop.

To move or copy a task, follow these steps:

1. Select the entire task by clicking the ID number for the task.
2. Click the Cut or Copy button. If you click Cut, the task should be removed from the list.
3. Click in the row where you want the task to appear.
4. Click the paste button.

To drag and drop a task, follow these steps:

a. Select the entire task by clicking the row header (ID number) for the task.
b. Point to the row header, or the top or bottom border of the selected area, with the arrowhead pointer.

c. Drag the mouse and when the shadowed I-beam appears just below where you want the task to appear, release the mouse button.

You can select multiple consecutive rows by dragging over the row headers with the four- headed cross. To select rows that are not consecutive, hold down the Ctrl key while clicking the row headers. You cannot drag and drop multiple nonconsecutive rows. You must cut and paste to move them. When you paste nonconsecutive rows into a new location, the rows become consecutive.

Inserting and Deleting Tasks

To insert a row, select the row below where you want the new row by clicking the row number choose Insert → New Task, or right click and choose New Task from the shortcut menu. If you want to insert multiple rows, select the number of rows you want to insert by clicking the first row number and dragging to select additional rows.

To delete entire tasks, select the row or rows you want to delete, and choose Edit→ Delete Task or press the Delete key on the keyboard.

You can also choose Edit→ Clear to delete tasks or parts of tasks. The Clear menu has options for clearing formats, contents, notes, hyperlinks, or the entire task.

Entering Recurring Tasks

When planning a project, it is not uncommon to plan for tasks that recur throughout the life of the project. A project review meeting is the most obvious example. If a task occurs more than once during the life of the project but does not occur at regular intervals, then you need to enter the task multiple times. However, if a task does recur at regular intervals, you can enter the task once and project can show it as a recurring task.

To enter a recurring task, follow these steps:

1. Click in the row where you want the recurring task to appear.
2. Choose Insert→ Recurring Task (Project inserts a row for you so there is no need to insert a blank row first).
3. Enter a Task Name in the recurring Task Information dialog box that appears
4. Enter a duration and recurrence pattern for one occurrence of the task. The Project Review meeting will be held for one hour every Thursday. Depending on whether you choose Daily, Weekly, Monthly, or Yearly, you have different options for the recurrence pattern.
5. Enter a date in the Start field for Range of Recurrence. If the task is an event, such as a meeting, enter the start time also.

6. Indicate whether you want the task to end after a specific number of occurrences or by a certain date.
7. Assign a calendar for scheduling the task, if you want. If you don't assign a calendar, Project uses the default standard calendar.
8. Click ok to set the recurrence pattern.

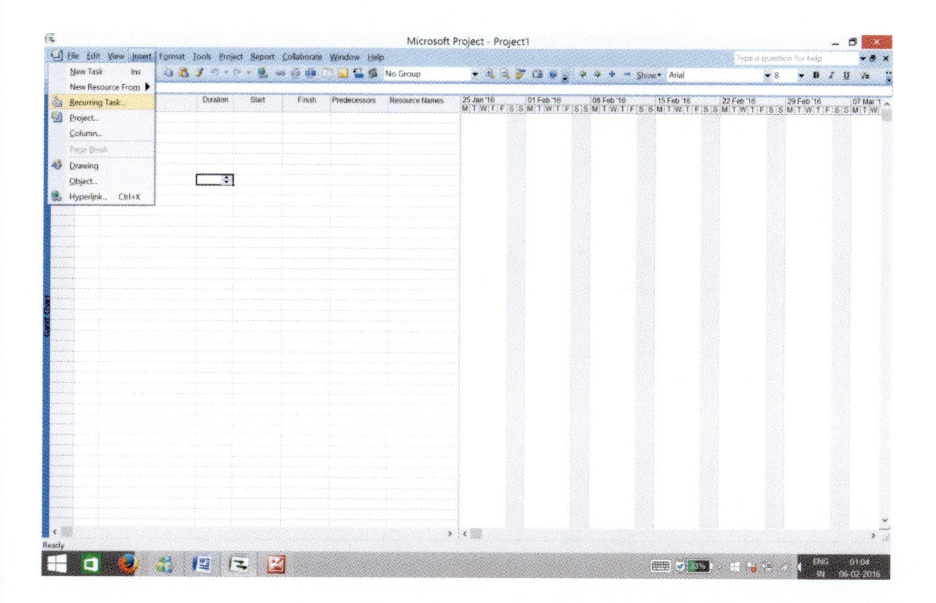

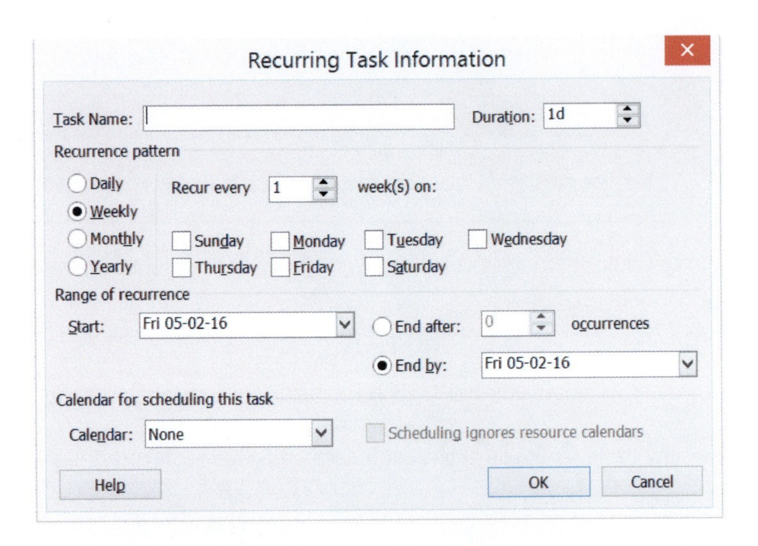

Grouping Tasks in Logical Order

Outlining helps organize your tasks into more manageable chunks. You can indent related tasks under a more general task, creating a hierarchy. The general tasks are called summary tasks; the indented tasks below the summary task are sub-tasks. A summary task's start and finish dates are determined by the start and finish dates of its earliest and latest subtasks.

1. Click once on the first activity of the group of activities you want to group. For the example Activities 4 and 5

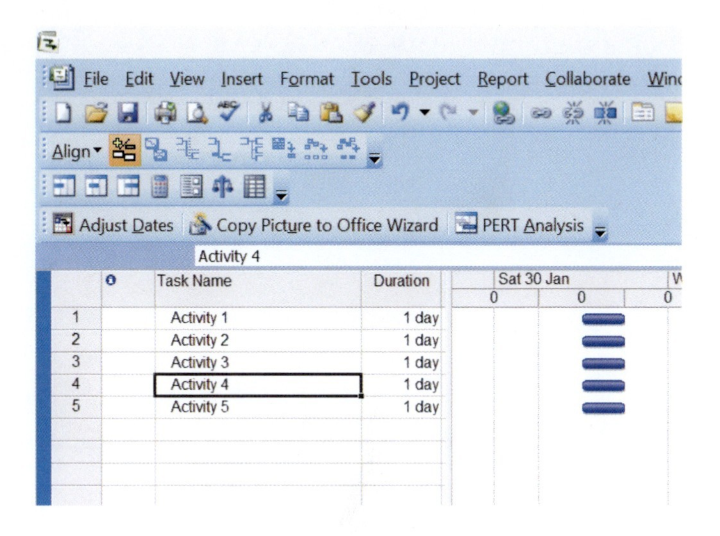

2. Then click on the option "New Task" in the "Insert" Menu to insert a new task that will represent the name of the group ("Group 1" for this example)

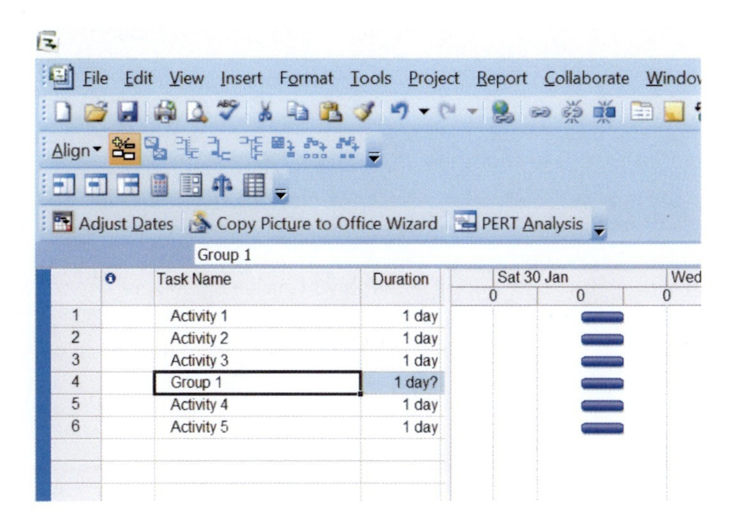

3. Then select the tasks below (4 and 5) and then click in the option "Outline-Indent" in the "Project" Menu

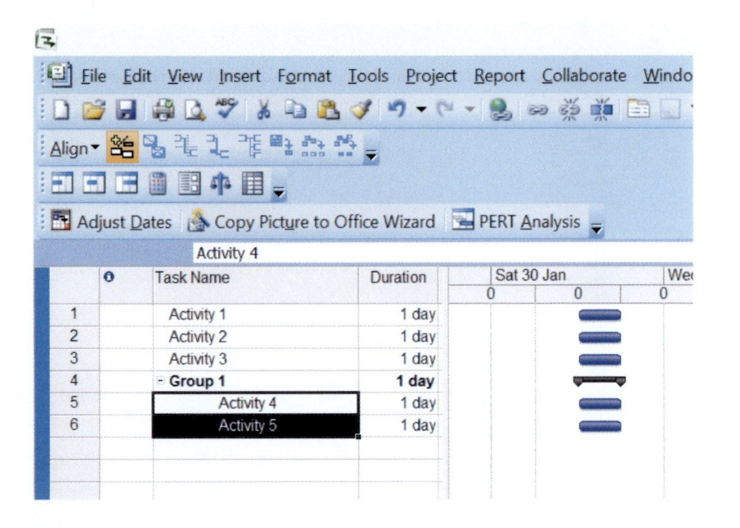

Creating Relationships between Tasks

A network of tasks in a project must be connecting activities from the start to the end, to establish these relationship we need to use the field "Predecessors" of each task, where we can designate which activity will be preceding the one we are updating, in the example below we will indicate MS project that "Activity 5" can start once "Activity 4" is completed (Finish to Start relationship).

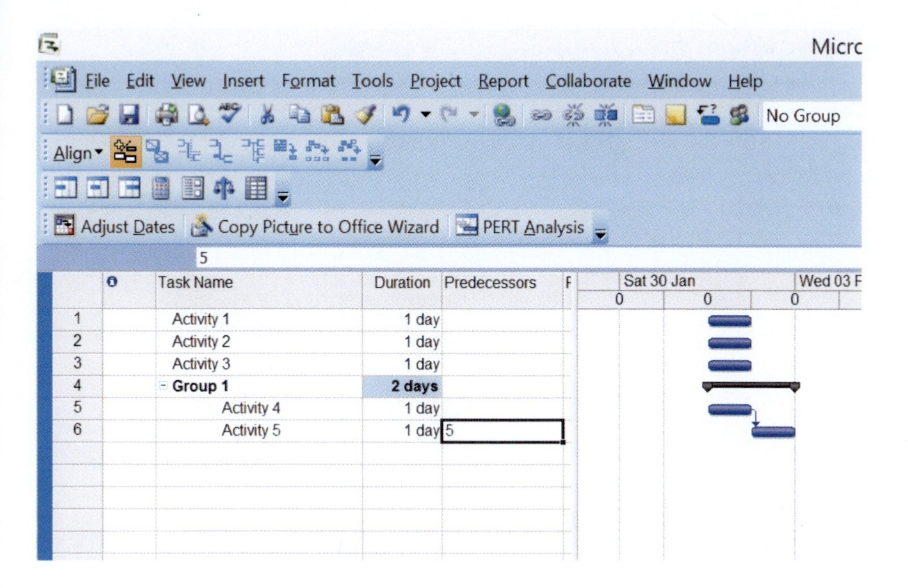

Notice that by establishing the relationship now the Group 1 takes 2 days to be completed, because before, the activities were set to be performed in parallel, and now they are in series (Finish to Start relationship)

Note: MS project will calculate dates based on the durations of the tasks, their relationships and the start date set for the project, however it is possible to change the starting date of a task (if necessary) By double clicking on a Task or milestone, and using the fields related to the dates (Start or Finish)

Assigning Resources to Tasks

You can use the Resource Sheet in Microsoft Project to create a list of the people, equipment, and material resources that make up your team and carry out the project tasks. Your resource list will consist of work resources or material resources. Work resources are people or equipment; material resources are consumable materials or supplies, such as concrete, wood, or nails.

1. On the View menu, click Resource Sheet.
2. On the View menu, point to Table, and then click Entry.
3. In the Resource Name field, type a resource name.
4. You can go through the fields in the sheet, but for the simplicity of the example just focus on the name and initials of the Resource
5. Below is an example of some Human resources added to the Resource Sheet (We could add also other type of resources such as Equipments, Consumables, etc.)

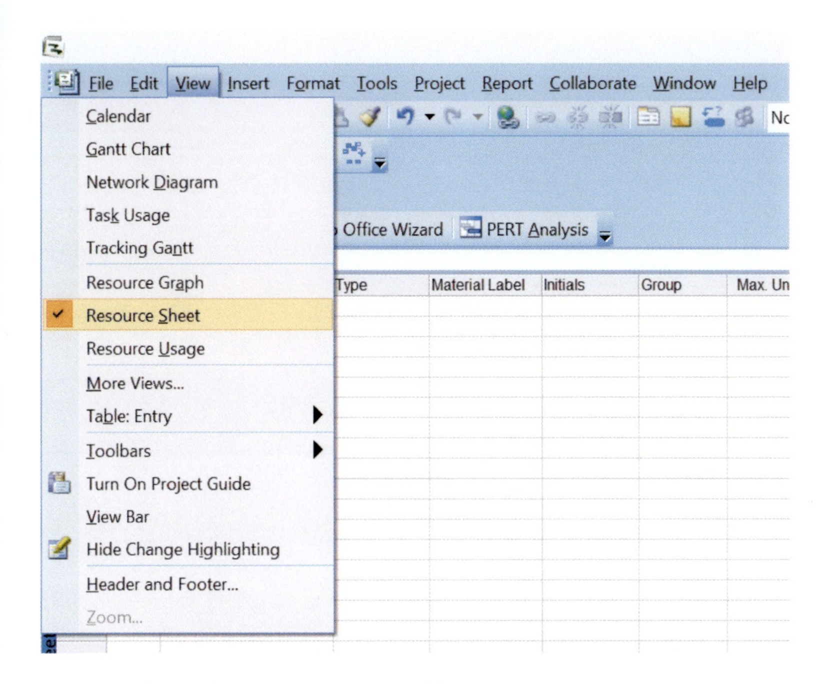
File
Edit
View
Insert
Format
Tools
Project
Report
Collaborate
Window
Help
Calendar
Gantt Chart
Network Diagram
Task Usage
Tracking Gantt
Resource Graph
Resource Sheet
Resource Usage
More Views...
Table: Entry
Toolbars
Turn On Project Guide
View Bar
Hide Change Highlighting
Header and Footer...
Zoom...
Office Wizard
PERT Analysis
Type
Material Label
Initials
Group
Max. Un

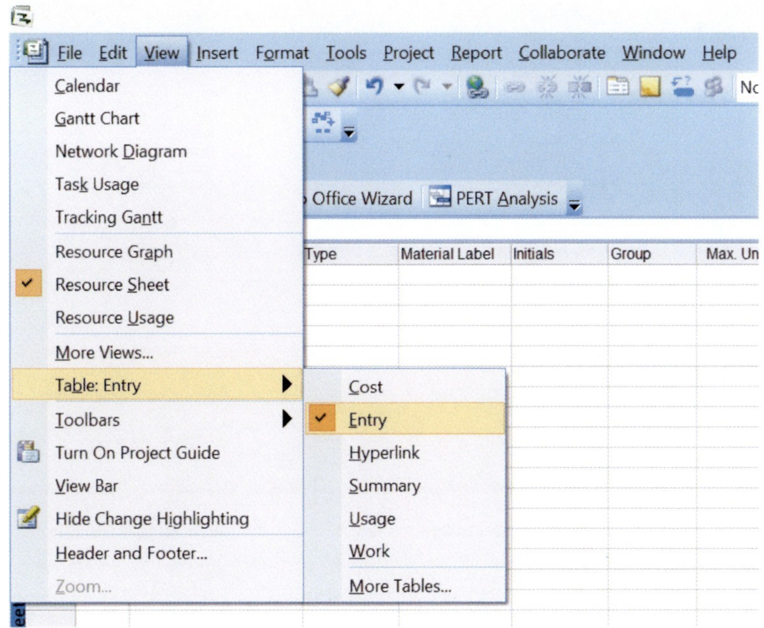
File
Edit
View
Insert
Format
Tools
Project
Report
Collaborate
Window
Help
Calendar
Gantt Chart
Network Diagram
Task Usage
Tracking Gantt
Resource Graph
Resource Sheet
Resource Usage
More Views...
Table: Entry
Toolbars
Turn On Project Guide
View Bar
Hide Change Highlighting
Header and Footer...
Zoom...
Cost
Entry
Hyperlink
Summary
Usage
Work
More Tables...
Office Wizard
PERT Analysis
Type
Material Label
Initials
Group
Max. Un

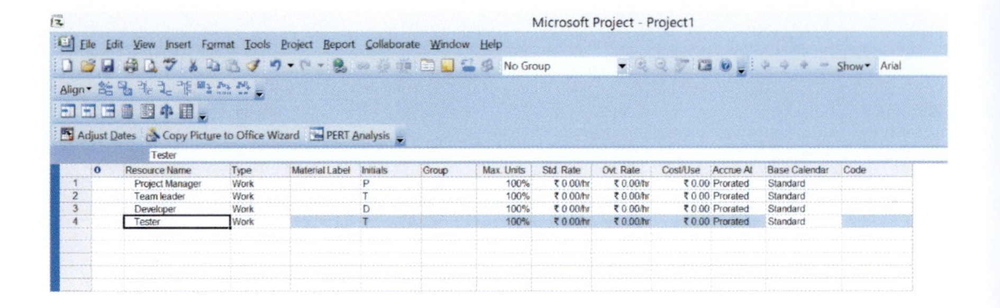

6. Once the resources are created, you can go back to the View menu, and click Gantt Chart to see again the tasks, and then when you double click a task you can add a resource to this task by using the tab "Resources"

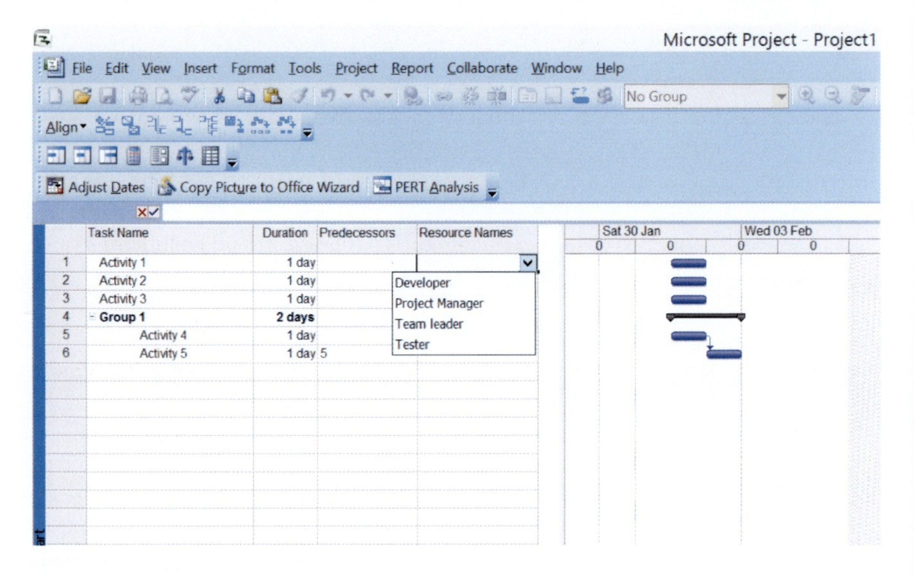

Note: The main goal of the resource assignment is to allocate properly the resources and to provide valuable information regarding the effort of the team.

PART-8

ESTIMATING PROJECT COST

Estimating Project Costs

In MS Project, you assign activity costs indirectly. Costs are assigned to resources. Examples of resources include the following:

- Employees at their hourly rate or prorated salaried rate, optionally including employee benefits
- Contractors
- Temporary employees
- Equipment at a lease rate or calculated periodic cost
- Facilities

Begin by creating a resource pool that includes the project resources. Resources from the pool are assigned to tasks, and the cost of a task is the cost of the resource multiplied by the amount of the resource used to complete the task.

If your colleagues use Project, you can use resources from a pool in another Microsoft Project file. This is more than a convenience- if the projects share staff, using a common resource pool helps ensure that staff members aren't accidentally overworked.

PART-9

CRITICAL PATH

Find Critical Path

- Critical Path Analysis (CPA) helps you to lay out all tasks that must be completed as part of a project.
- CPA helps you to identify the minimum length of time needed to complete a project
- For finding CP list all the activities and enter early start, late start, early finish and late finish info of all the activities.
- You can do this under insert/columns and selecting each terms.
- Following screen shot demonstrates how to insert.
- Project automatically calculates ES, EF, LS and LF based on the starting/ending dates you have provided.

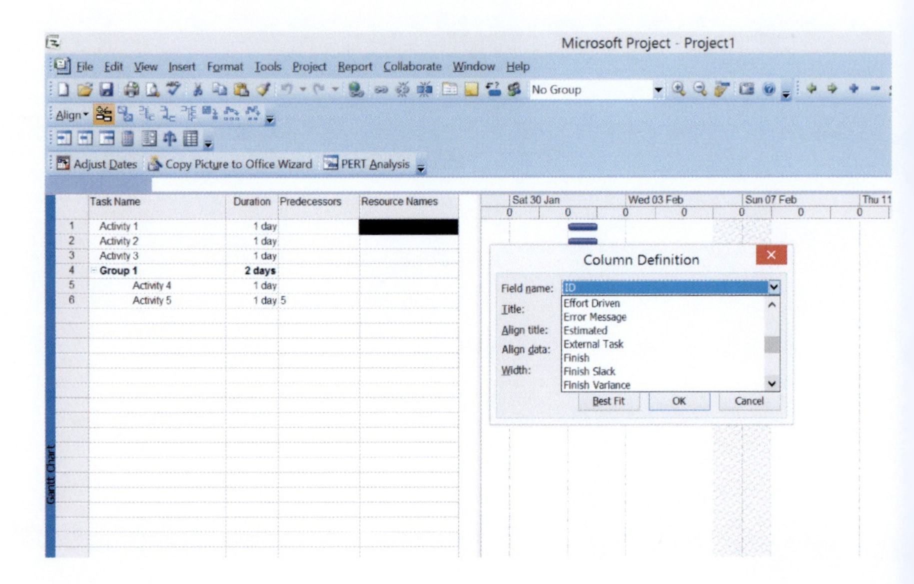

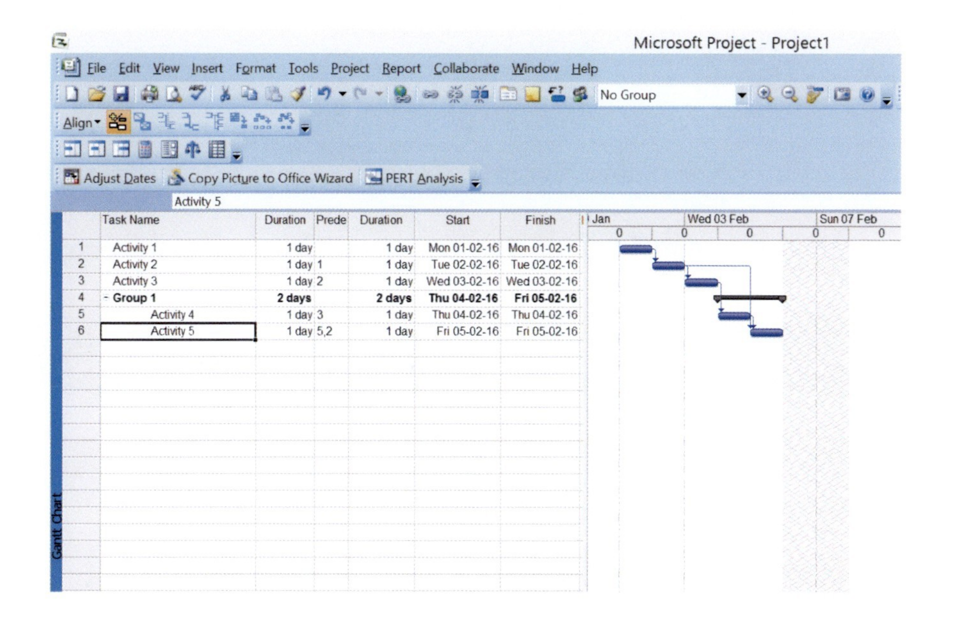

Showing Critical Path

1. You can show the critical path in multiple ways. You can show critical path using Gantt chart, Detailed Gantt, Network Diagram, and showing only critical tasks.
2. In Gantt Chart click on Gantt Chart wizard and choose critical path and follow the steps.
3. In Detailed Gantt click on View →More Views →Detailed Gantt →Apply. It shows the critical path with slack time.

4. In Network Diagram click on View →More Views →Network Diagram →Apply.
5. For showing only the critical tasks, click on Gantt Chart →filter →Critical.
6. The following slides have screenshots of how to show critical path using various methods.

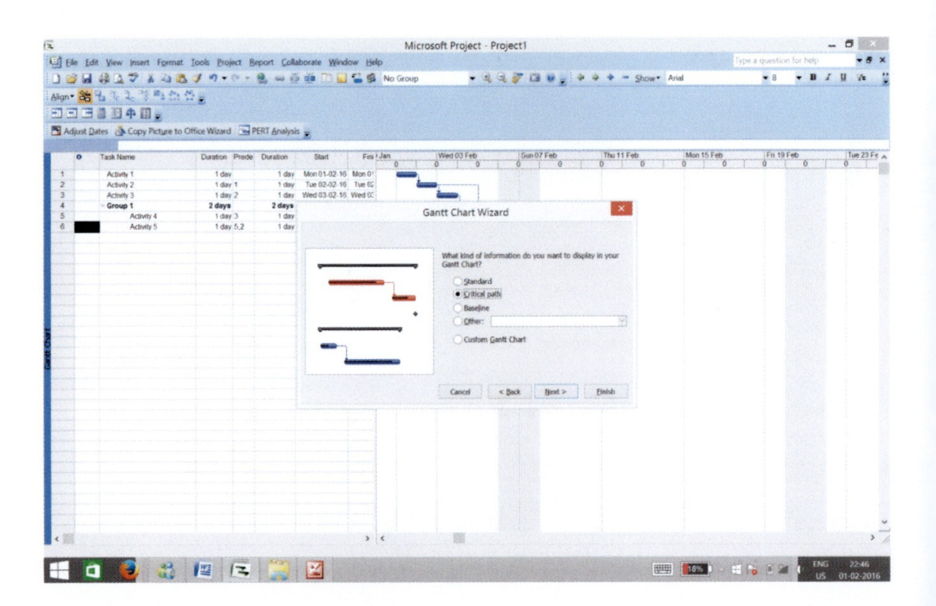

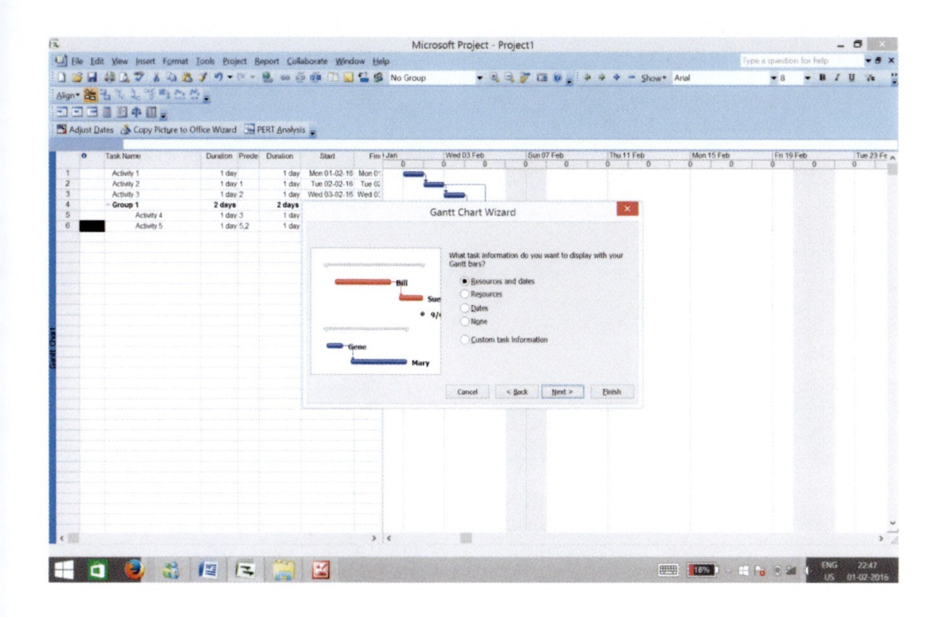
Microsoft Project - Project1
Gantt Chart Wizard
What task information do you want to display with your Gantt bars?
Resources and dates
Resources
Dates
None
Custom task information
Cancel
< Back
Next >
Finish

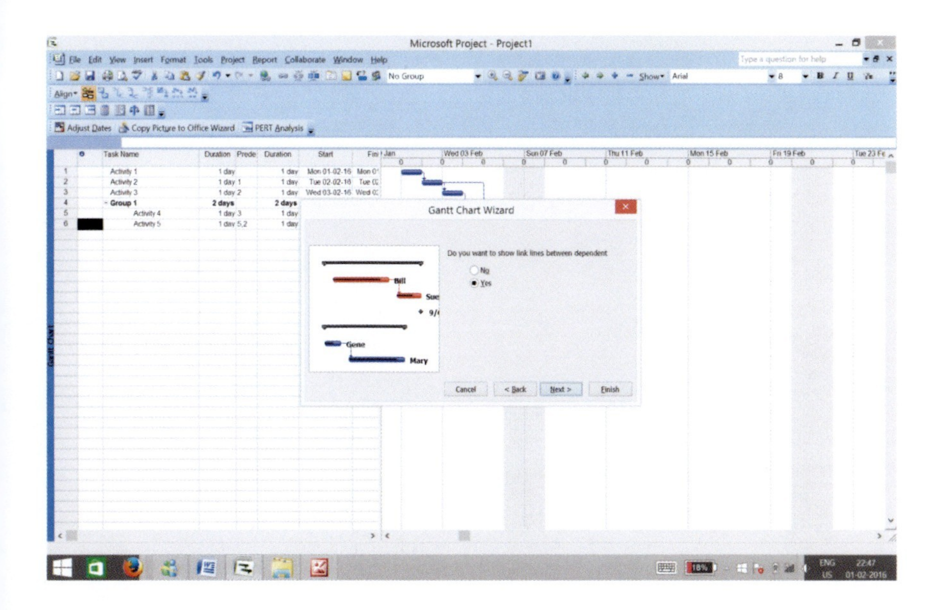
Microsoft Project - Project1
Gantt Chart Wizard
Do you want to show link lines between dependent
No
Yes
Cancel
< Back
Next >
Finish

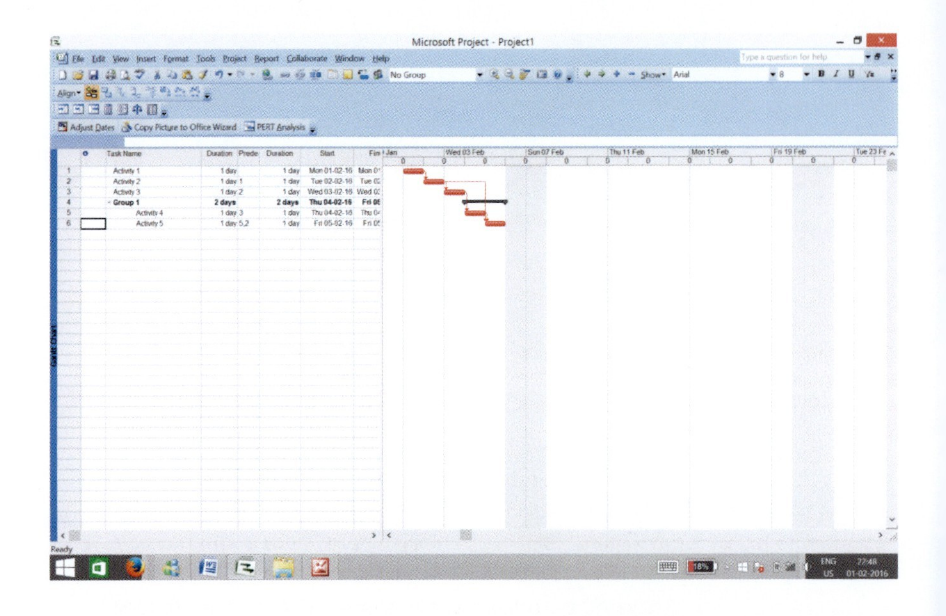
Microsoft Project - Project1
File Edit View Insert Format Tools Project Report Collaborate Window Help
Adjust Dates
Copy Picture to Office Wizard
PERT Analysis

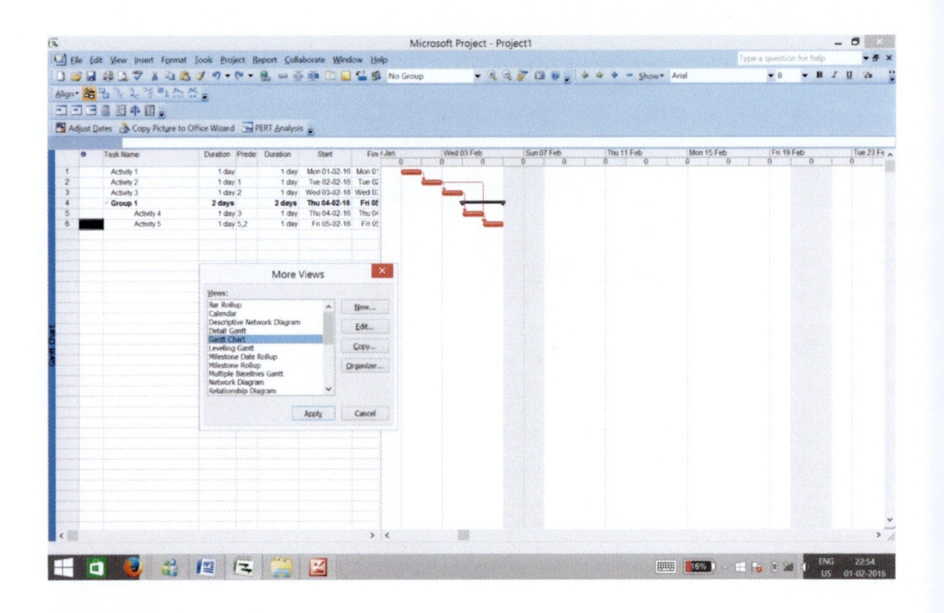
Microsoft Project - Project1
File Edit View Insert Format Tools Project Report Collaborate Window Help
Adjust Dates
Copy Picture to Office Wizard
PERT Analysis
More Views
Views:
Bar Rollup
Calendar
Descriptive Network Diagram
Detail Gantt
Gantt Chart
Leveling Gantt
Milestone Date Rollup
Milestone Rollup
Multiple Baselines Gantt
Network Diagram
Relationship Diagram
New...
Edit...
Copy...
Organizer...
Apply
Cancel

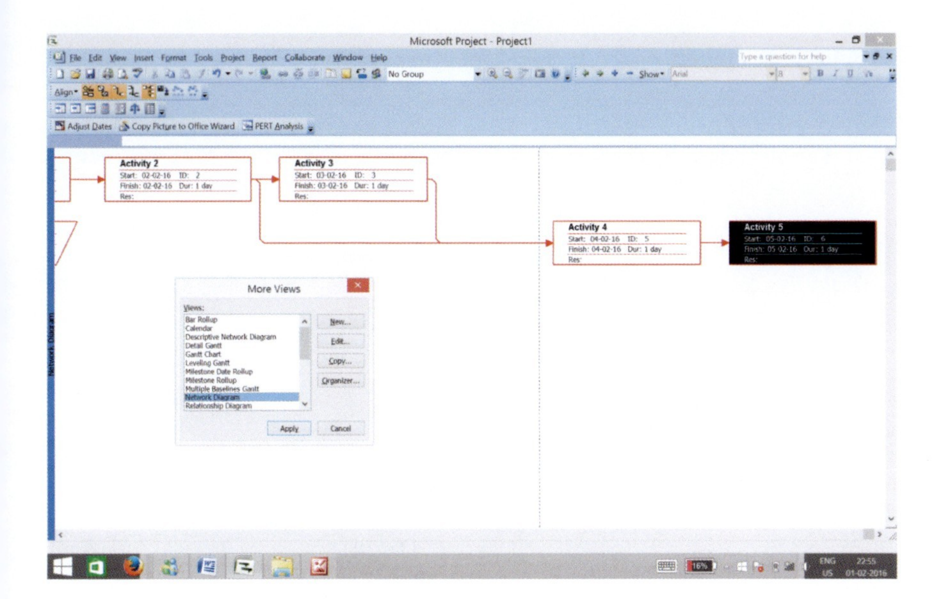

Note: Accordingly, require view can be selected from the provided dialogue box i.e. network diagram, leveling gantt, detail gantt etc.

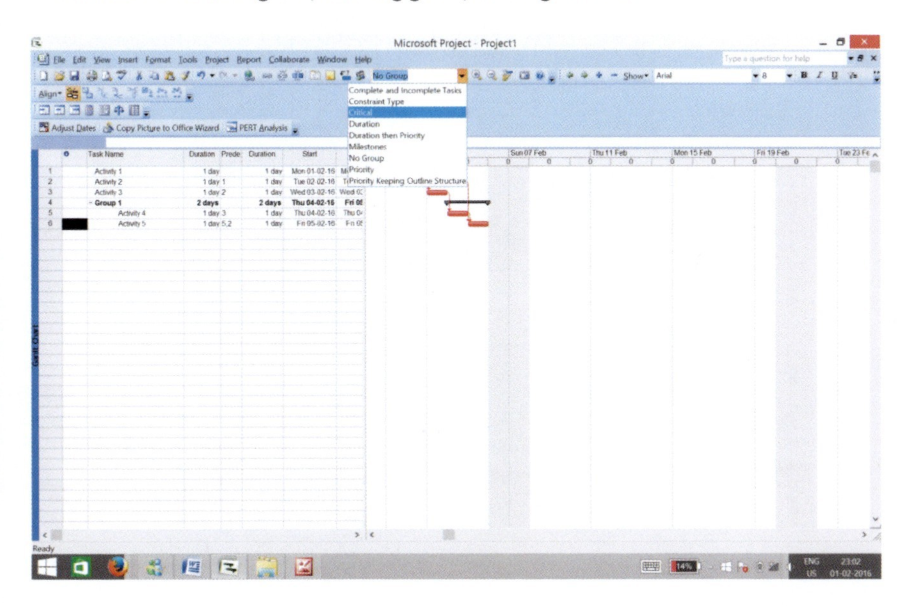

Slack Time for Scheduling

1. For viewing the schedule showing the slack go to Views →More views → Detailed Gantt view →Apply.
2. In this click on View →Table →Schedule
3. Slack appears as thin bars to the right of a task, with slack values adjoining the regular Gantt bars
4. You can also view the free slack and total slack of a task in the sheet.
5. You can move the activity within the available slack time, to balance the resources, in the cases where over allocation is present.

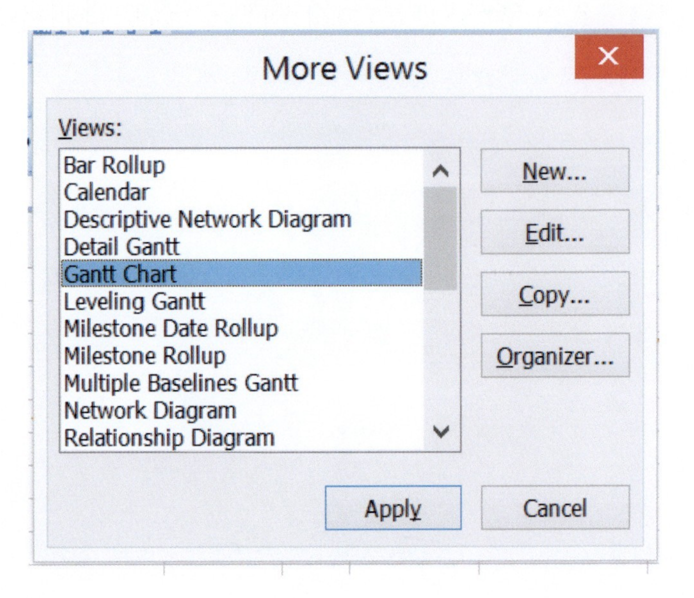

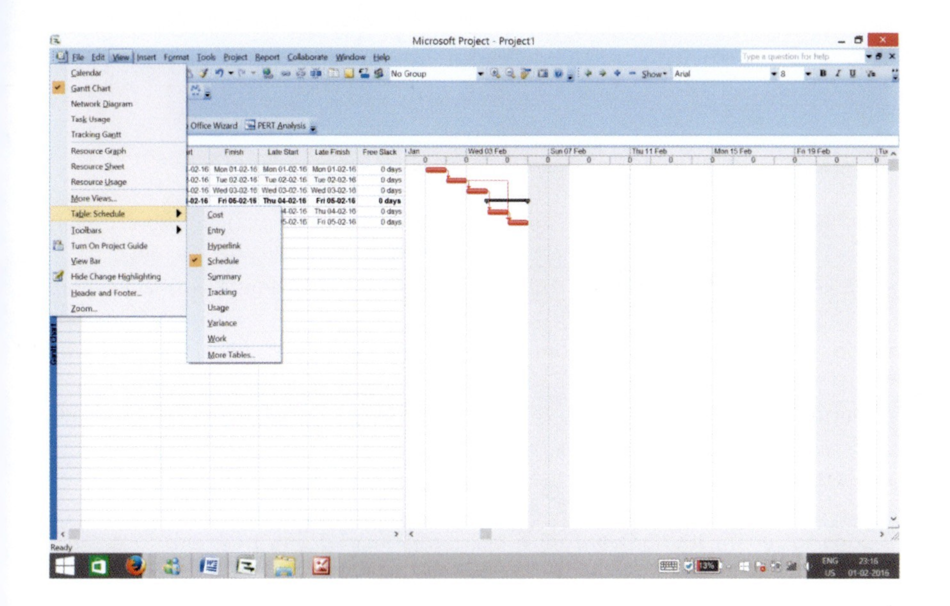

Microsoft Project - Project1
File Edit View Insert Format Tools Project Report Collaborate Window Help
Calendar
Gantt Chart
Network Diagram
Task Usage
Tracking Gantt
Resource Graph
Resource Sheet
Resource Usage
More Views...
Table: Schedule
Toolbars
Turn On Project Guide
View Bar
Hide Change Highlighting
Header and Footer...
Zoom...
Cost
Entry
Hyperlink
Schedule
Summary
Tracking
Usage
Variance
Work
More Tables...
Office Wizard
PERT Analysis
Finish
Late Start
Late Finish
Free Slack
Ready

PART-10

PRINTING OF THE PROJECT

Printing Views

To print the Gantt Chart or any other view, first use the View menu to display the view. Choose the data you want to display by selecting a table (View→ Table), and then adjust the columns and position the timeline. The Gantt Chart prints with the timescale displayed. You may want to print a more condensed timescale, either because your project spans a number of months or the audience for the printout needs only summary data. To change the timescale, right- click the timescale at the top of the chart and choose Timescale from the shortcut menu, or choose Format→ Timescale from the menu. In the Timescale dialog box, set the Major and Minor scales (for example, weeks and days, days and hours, months and days). Click the preview in the dialog box as you change scales and set other display options. Click the nonworking time tab to change the way weekends and holidays are displayed in the view.

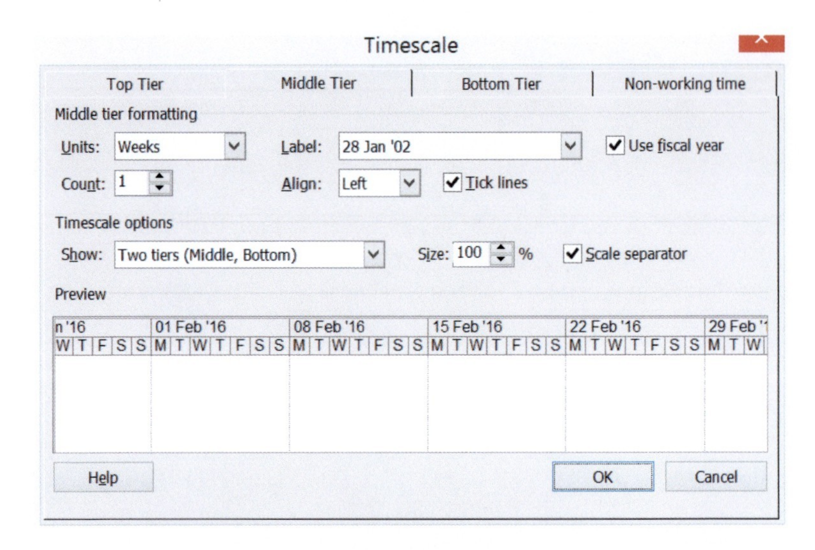

Click the print button on the standard toolbar to print the view as it appears. To set a range of dates or print multiple copies, choose File→ Print from the Menu bar to open the print dialog box.

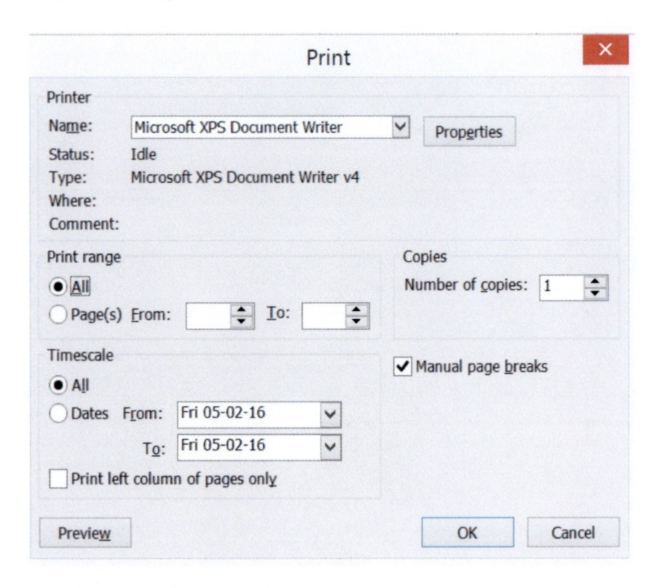

PART-11

SAVING OF THE PROJECT

Saving using auto save option

If you do not always remember to save your project files, you can have project automatically save your project files at regular intervals. You can find the Auto Save option on the Save tab in of the Options dialog box (Tools→ Options), shown in the figure. You can decide how often you want the project to be saved and whether you want to save just the active project or all open project files. You can also specify whether or not you want to be prompted before it saves.

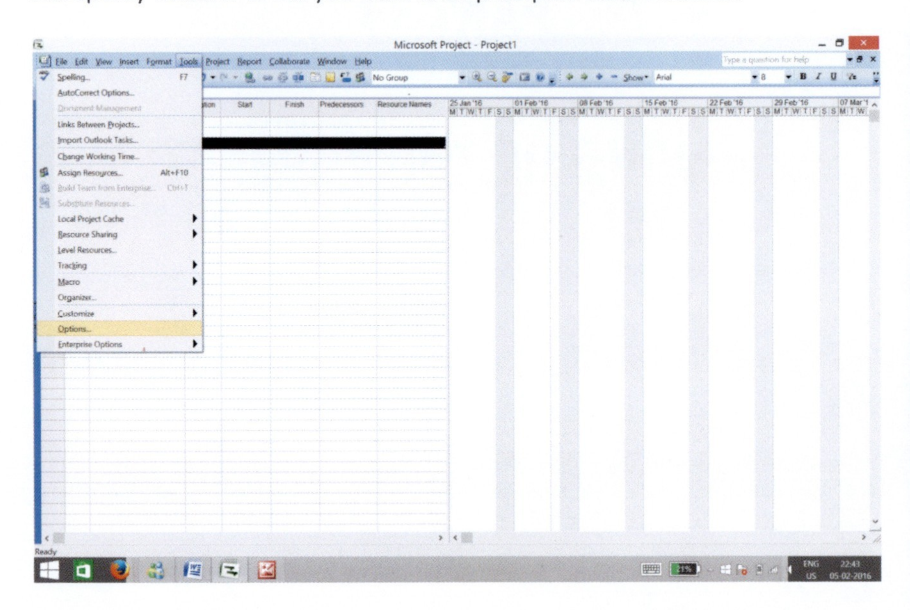

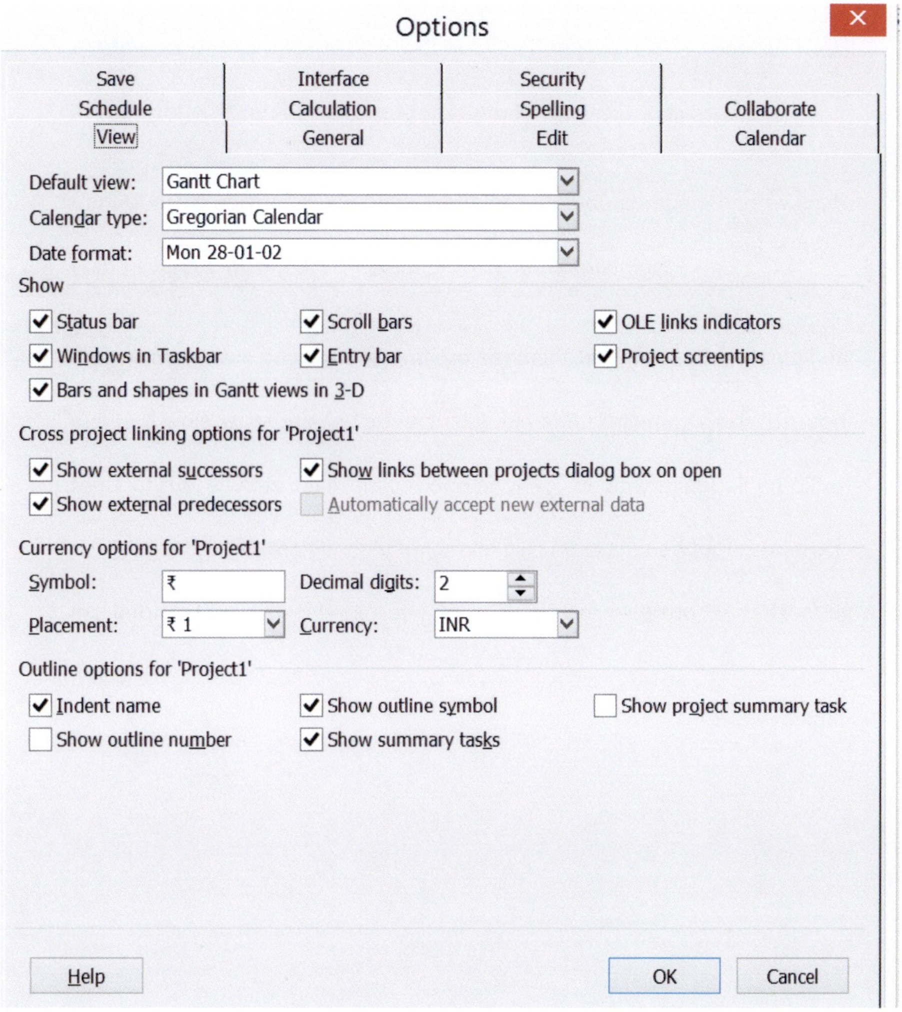

Options

Save | Interface | Security
Schedule | Calculation | Spelling | Collaborate
View | General | Edit | Calendar

Default view: Gantt Chart

Calendar type: Gregorian Calendar

Date format: Mon 28-01-02

Show

- [x] Status bar
- [x] Scroll bars
- [x] OLE links indicators
- [x] Windows in Taskbar
- [x] Entry bar
- [x] Project screentips
- [x] Bars and shapes in Gantt views in 3-D

Cross project linking options for 'Project1'

- [x] Show external successors
- [x] Show links between projects dialog box on open
- [x] Show external predecessors
- [ ] Automatically accept new external data

Currency options for 'Project1'

Symbol: ₹ Decimal digits: 2

Placement: ₹ 1 Currency: INR

Outline options for 'Project1'

- [x] Indent name
- [x] Show outline symbol
- [ ] Show project summary task
- [ ] Show outline number
- [x] Show summary tasks

Help OK Cancel

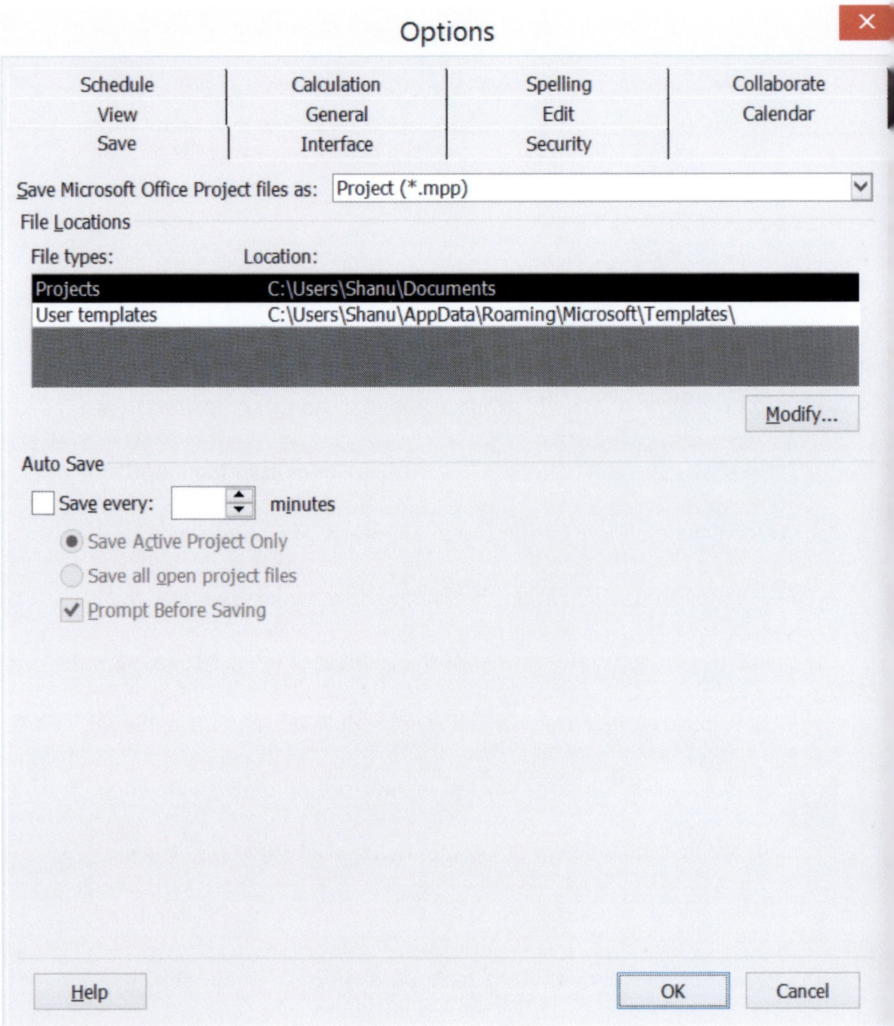

Options

Schedule	Calculation	Spelling	Collaborate
View	General	Edit	Calendar
Save	Interface	Security	

Save Microsoft Office Project files as: Project (*.mpp)

File Locations

File types:	Location:
Projects	C:\Users\Shanu\Documents
User templates	C:\Users\Shanu\AppData\Roaming\Microsoft\Templates\

Modify...

Auto Save

Save every: minutes

Save Active Project Only

Save all open project files

Prompt Before Saving

Help OK Cancel

PART-12

LIVE PROJECT

Let try an example

It's a actual example taken from the project of a construction company showing their all related construction activities involved during the whole process.

Now see what we have study till in MS Project in the form of an example.

1. Start with a new project name "Project 1"

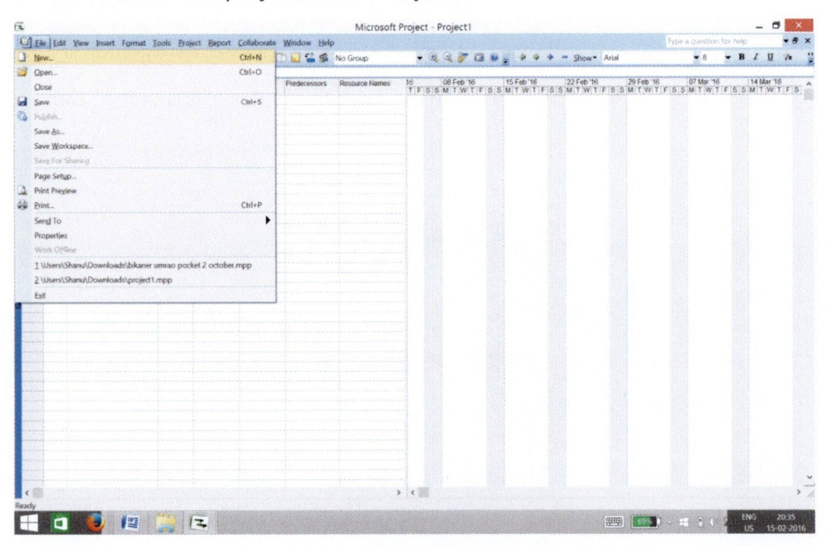

2. Assign the start date of the project to be Feb-15-2016 in project information menu.

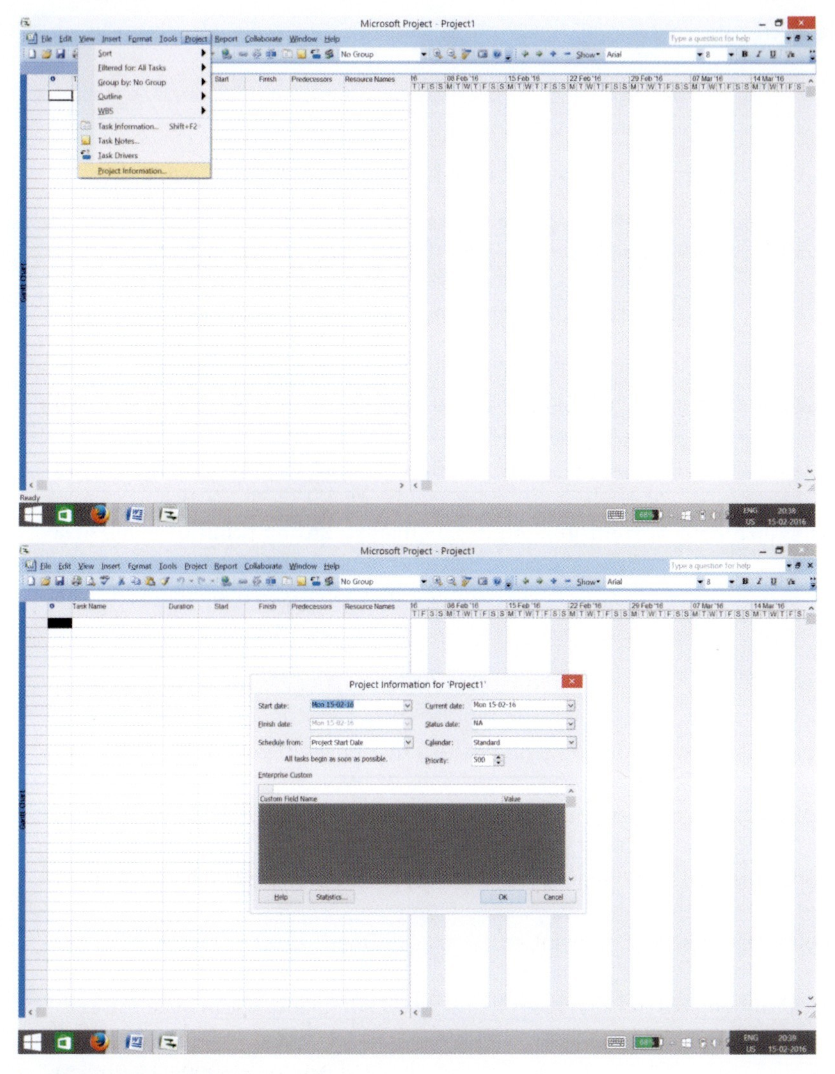

3. Now adding tasks

a) Write the name of each task in the spreadsheet using the column "Task Name".

b) Write the duration in days of each task in the spreadsheet using the column "Duration".

c) Group the tasks by the Phase according to the table of tasks shown before, and add a group that encloses the phases named “Brick work” this will represent the plan as a whole.

d) Write the predecessors of each task in the spreadsheet using the Column “Predecessors” (If you can’t see the column, try to expand the vertical bar that divides the spreadsheet to the Gantt Chart).

e) To convert a Task in a Milestone, just double click the Task and go to the tab “Advanced” then check the box that says “Mark Task as a Milestone”.

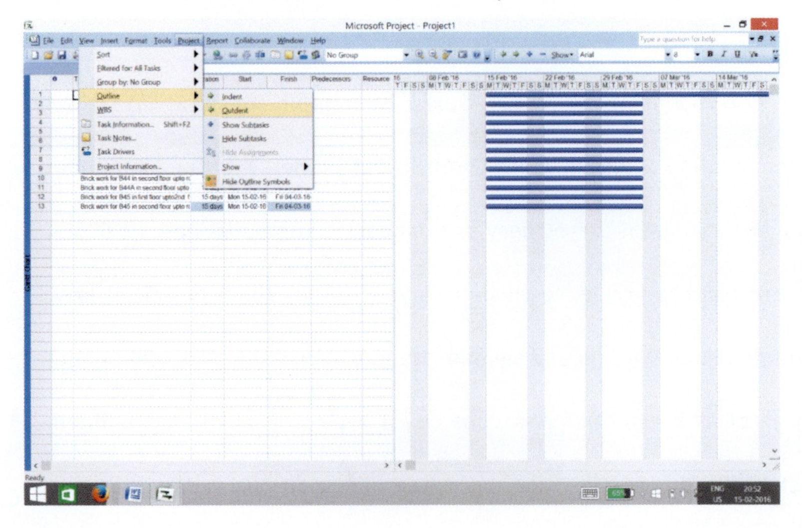

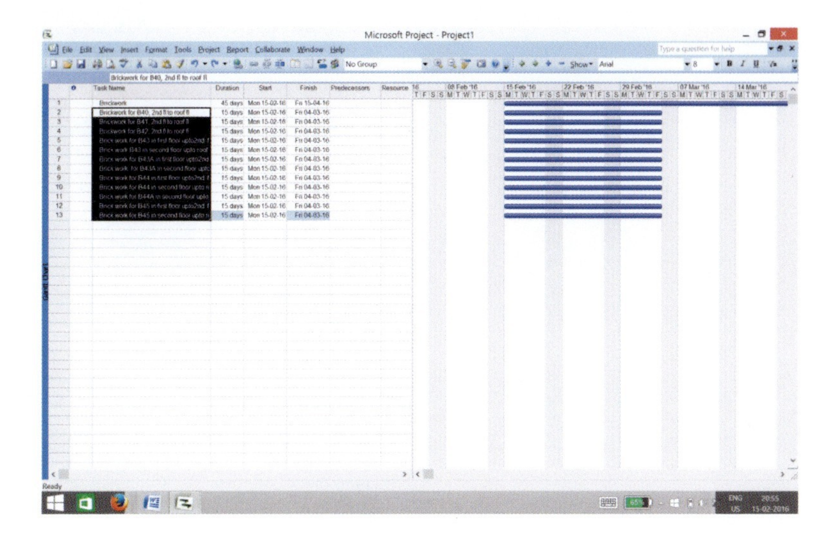

(Highlight the tasks that are going to be added as subtasks)

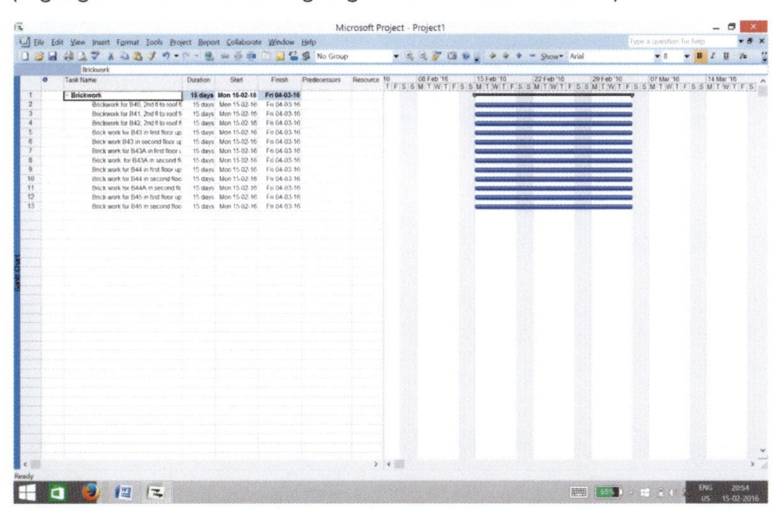

(The final result should look like this, now repeat this steps to create the Sub-groups that will represent the phases (Definition and Design))

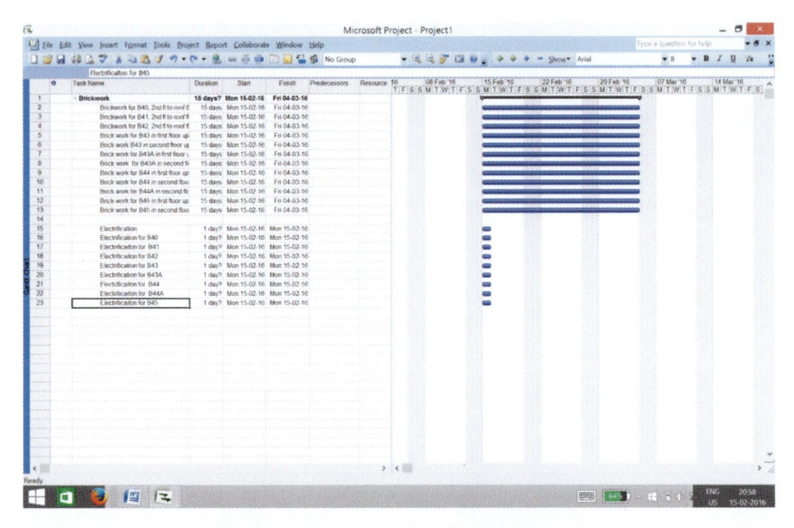

Insert a new task at the beginning of the definition tasks

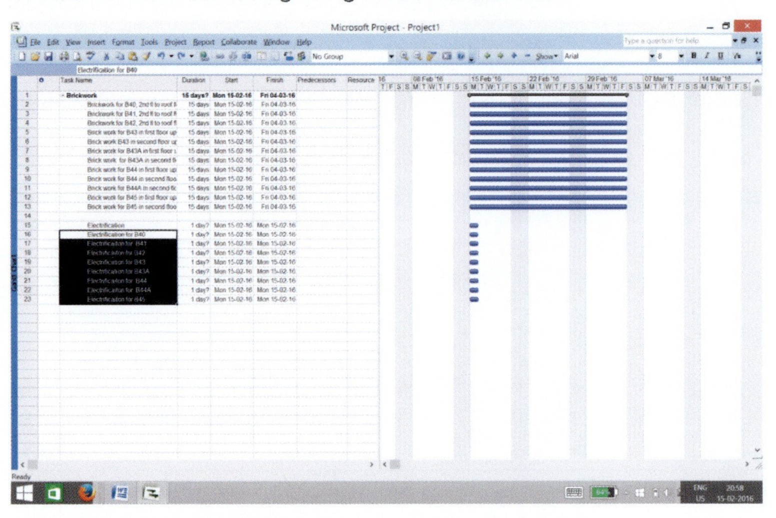

Highlight the tasks that are going to be added as subtasks

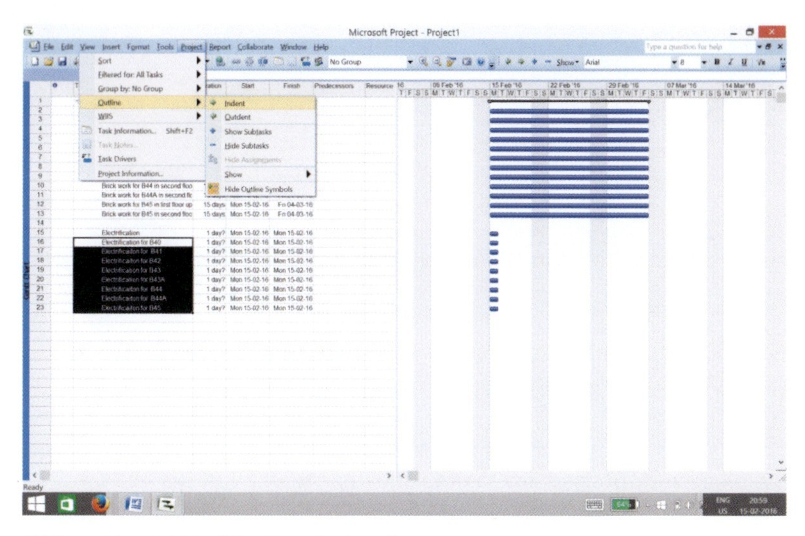

Click on the option “Outline -Indent”

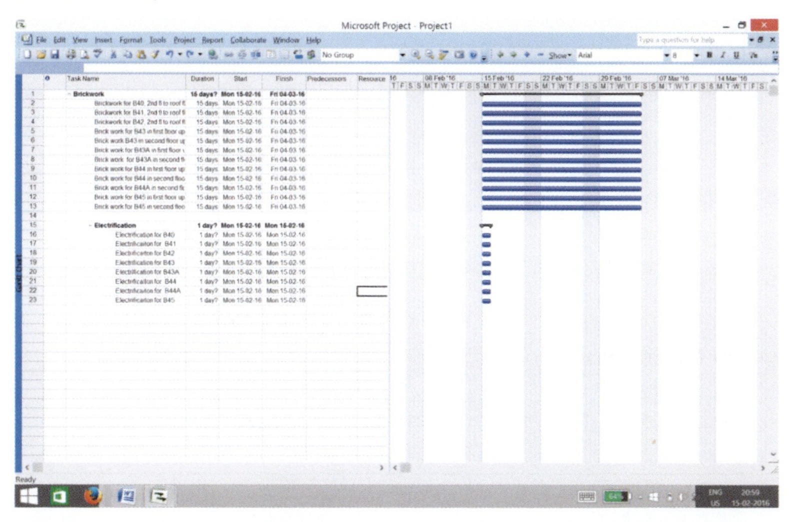

The final result should look like this, now repeat this steps to create the Sub-group that will represent the phase “Design”

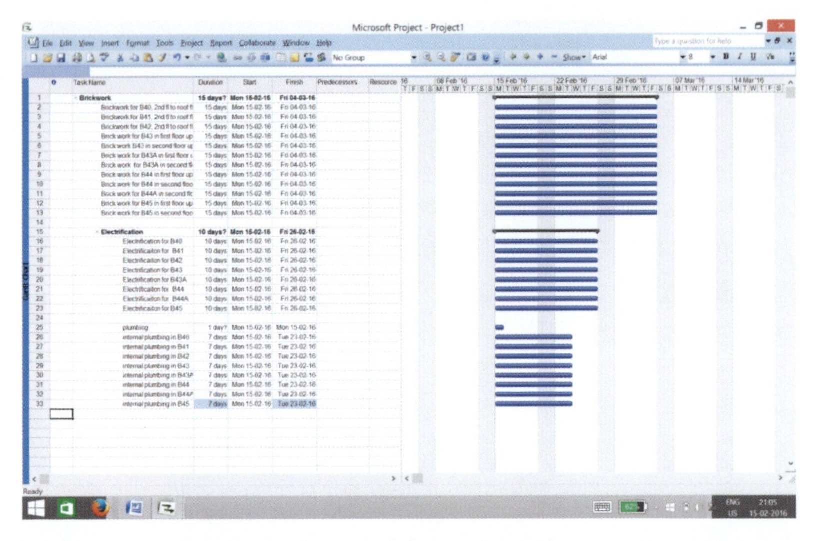

Insert a new task at the beginning of the Design tasks

(Notice that the new task that will work as a group for the "Design Phase" is inside the group "Definition Phase", therefore we need to Outdent one position to put it at the same level as the Definition Phase)

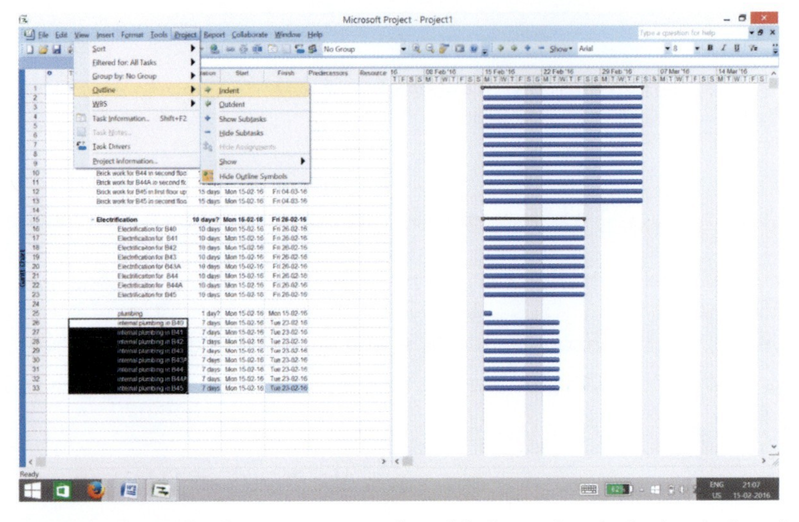

Highlight the tasks that are going to be added as subtasks in the design phase and then Click on the option "Outline -Indent"

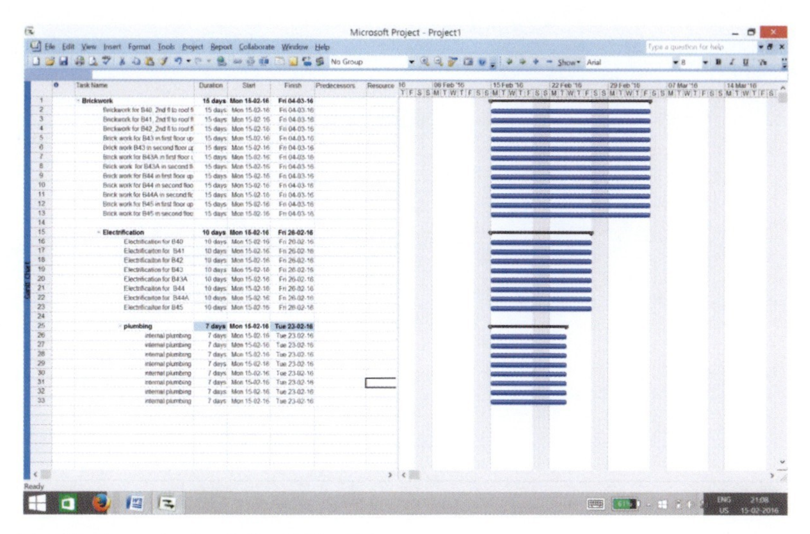

The final result should look like this.

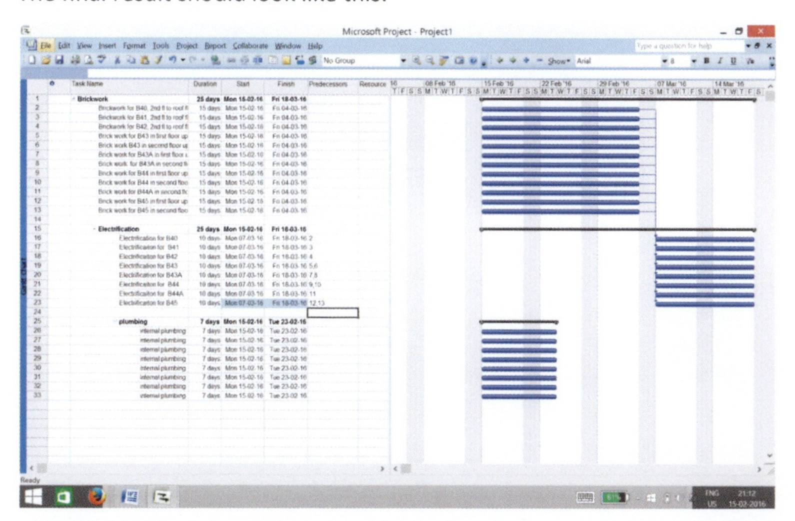

Linking all task accordingly to the Predecessors (Brickwork to Electrification to Plumbing).

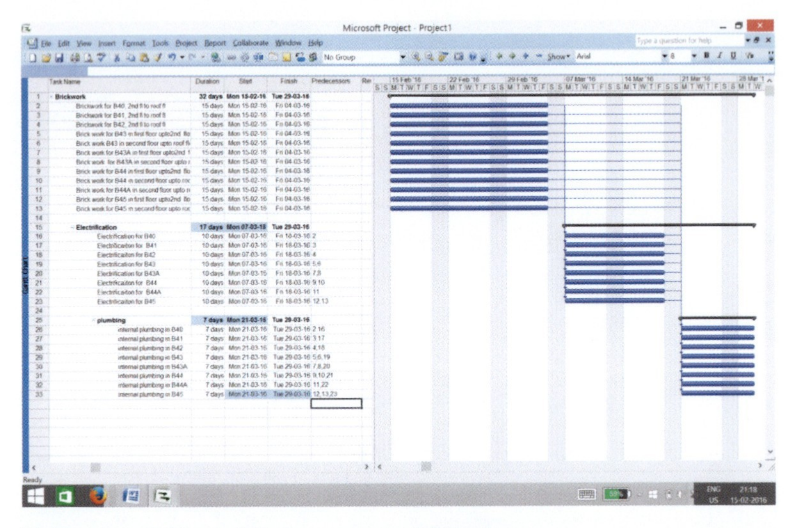

Making Gantt Chart clear in a proper visible size.

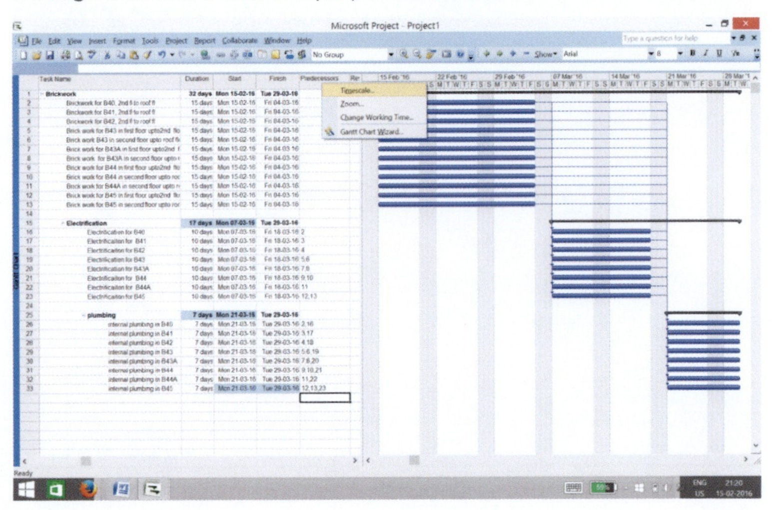

Making Gantt Chart clear in a proper visible size.

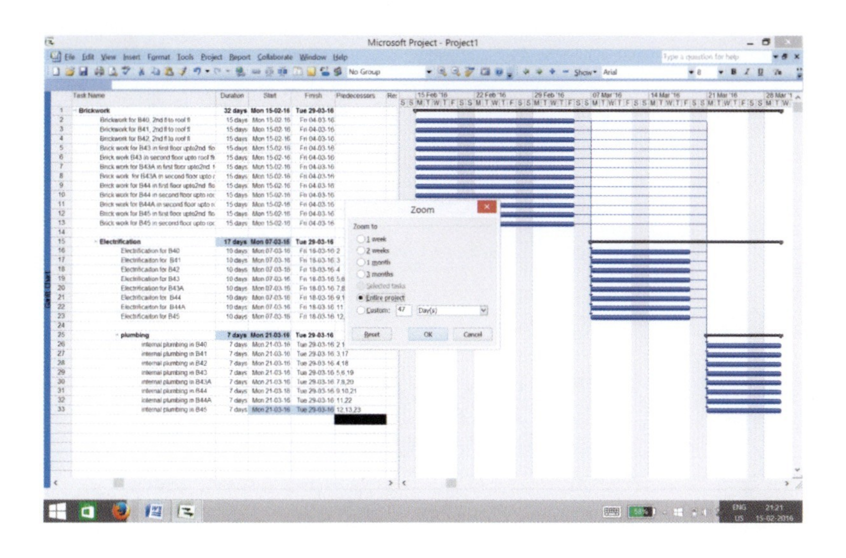
Microsoft Project - Project1
File Edit View Insert Format Tools Project Report Collaborate Window Help
No Group
Zoom
Zoom to
1 week
2 weeks
1 month
3 months
Selected tasks
Entire project
Custom:
Reset
OK
Cancel
Brickwork
Electrification
plumbing

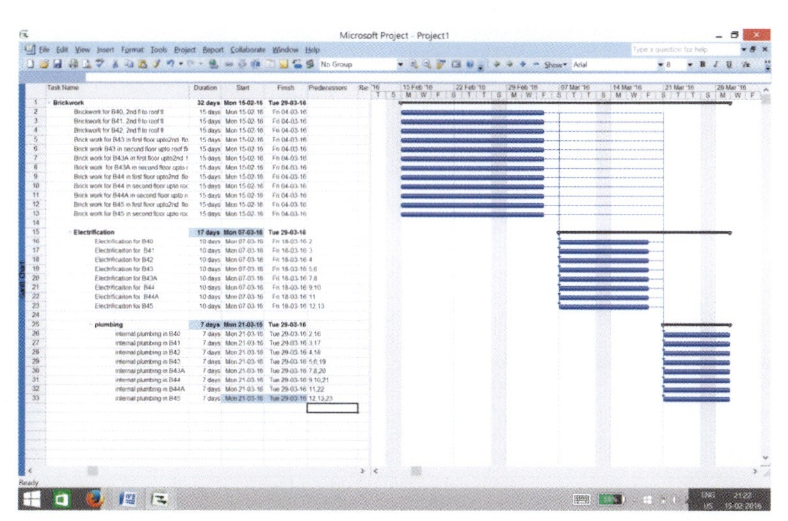
Microsoft Project - Project1
File Edit View Insert Format Tools Project Report Collaborate Window Help
No Group
Brickwork
Electrification
plumbing
Ready

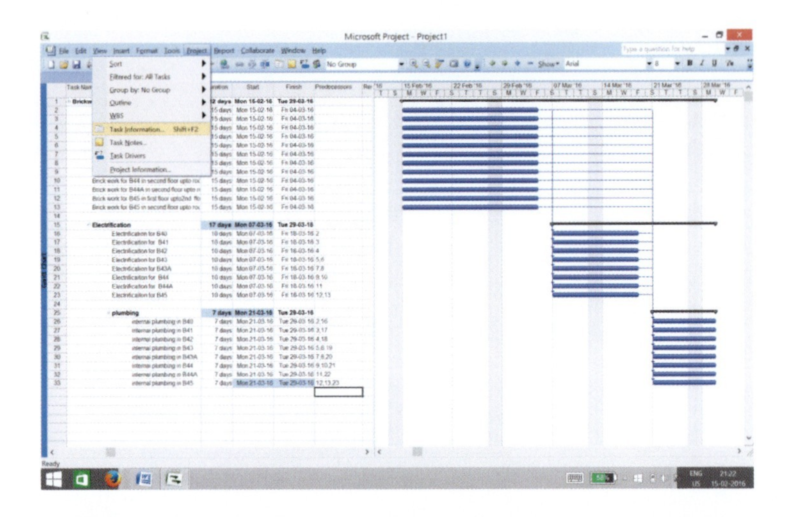
Microsoft Project - Project1
File
Edit
View
Insert
Format
Tools
Project
Report
Collaborate
Window
Help
Sort
Filtered for: All Tasks
Group by: No Group
Outline
WBS
Task Information... Shift+F2
Task Notes...
Task Drivers
Project Information...
Electrification
plumbing

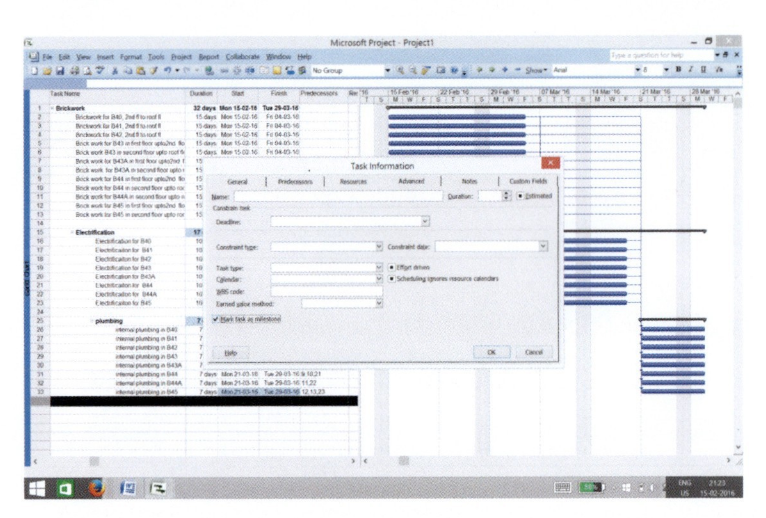
Microsoft Project - Project1
Brickwork
Electrification
plumbing
Task Information
General
Predecessors
Resources
Advanced
Notes
Custom Fields
Name:
Duration:
Estimated
Constrain task
Deadline:
Constraint type:
Constraint date:
Task type:
Effort driven
Calendar:
Scheduling ignores resource calendars
WBS code:
Earned value method:
Mark task as milestone
Help
OK
Cancel

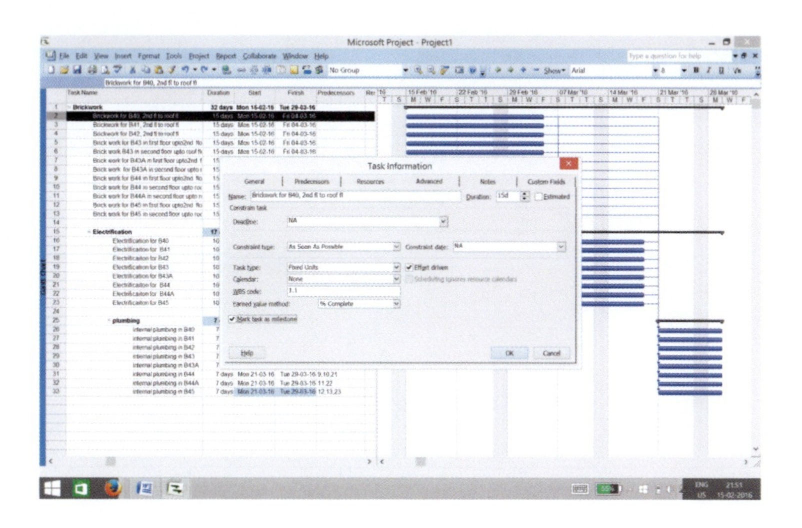

4. Now adding the resource
   a) Got to the view "Resource Sheet".
   b) Add the necessary resources to the "Resources Sheet", we are going to use only the Name, Initials and Standard Rate in ₹/hr. The resources are going to be taken from the table showed at the beginning of the example, more specifically from the column "Responsibilities".
   c) Now, with the Resources already register in the project file, go back to the View "Gantt Chart".

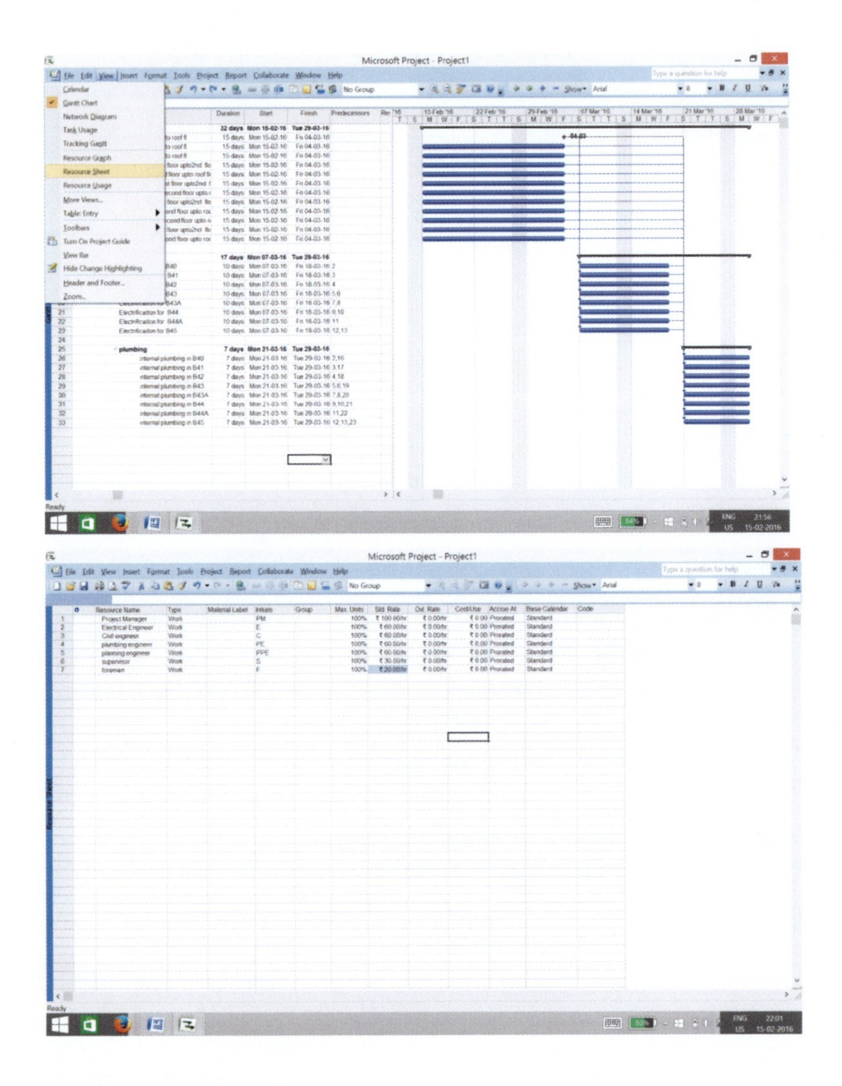

Microsoft Project - Project1
File Edit View Insert Format Tools Project Report Collaborate Window Help
Calendar
Gantt Chart
Network Diagram
Task Usage
Tracking Gantt
Resource Graph
Resource Sheet
Resource Usage
More Views...
Table: Entry
Toolbars
Turn On Project Guide
View Bar
Hide Change Highlighting
Header and Footer...
Zoom...
plumbing
Microsoft Project - Project1
File Edit View Insert Format Tools Project Report Collaborate Window Help
No Group
Resource Name Type Material Label Initials Group Max. Units Std. Rate Ovt. Rate Cost/Use Accrue At Base Calendar Code
1 Project Manager Work PM 100% ₹ 100.00/hr ₹ 0.00/hr ₹ 0.00 Prorated Standard
2 Electrical Engineer Work E 100% ₹ 60.00/hr ₹ 0.00/hr ₹ 0.00 Prorated Standard
3 Civil engineer Work C 100% ₹ 60.00/hr ₹ 0.00/hr ₹ 0.00 Prorated Standard
4 plumbing engineer Work PE 100% ₹ 60.00/hr ₹ 0.00/hr ₹ 0.00 Prorated Standard
5 planning engineer Work PPE 100% ₹ 60.00/hr ₹ 0.00/hr ₹ 0.00 Prorated Standard
6 supervisor Work S 100% ₹ 30.00/hr ₹ 0.00/hr ₹ 0.00 Prorated Standard
7 foreman Work F 100% ₹ 20.00/hr ₹ 0.00/hr ₹ 0.00 Prorated Standard
Ready

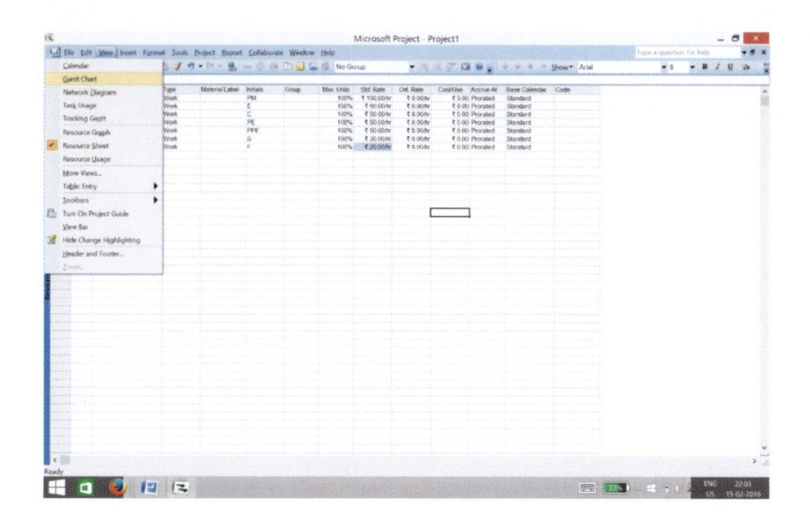

5. Assign the resources

a) Double click the task you want to link to resources available in the "Resource Sheet".

b) Then got to the Tab "Resources" and look up the resources you want to relate to the activity (For the example let's keep the amount of effort of each Resources as 100%, Leveling Resources won't be covered in this tutorial), finally Click the "Ok" button to finish the assignment.

c) Repeat steps 1 and 2 for the rest of the tasks.

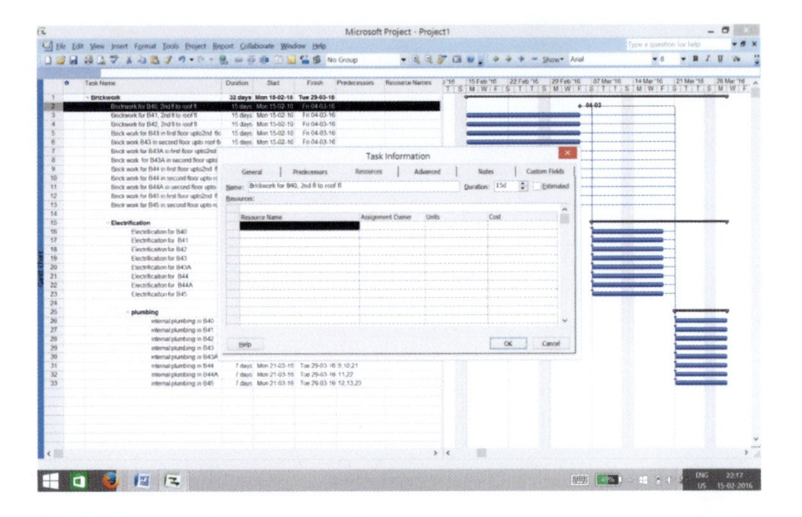

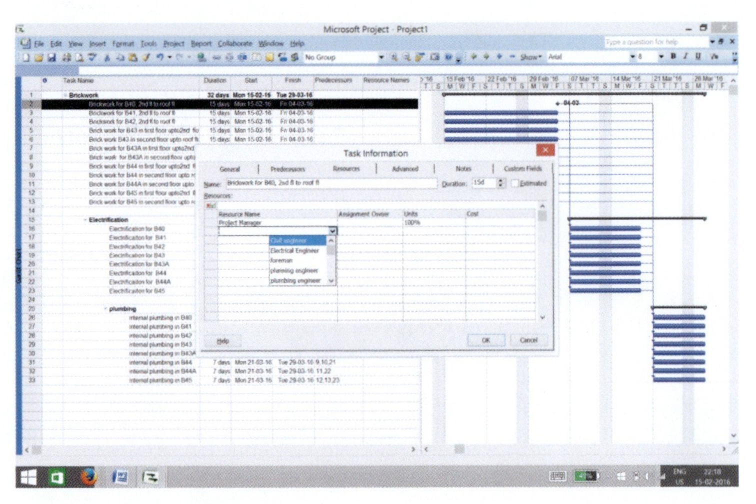

6. Adjust the Gantt chart

a) Adjust the length of the Gantt Chart such that it can be seen in one screen (If Possible), to do this perform a Right Click on top of the Gantt Chart first, a pop-up menu will appear, select the option “Networking Time…”

b) In the form that will open go to the tab “Time Scale” and Change the Major Scale Units to “Months” and the Minor Scale Units to “Weeks”, then press the “Ok” button to see the results in the Gantt Chart. (Adjust as necessary the scales once you are familiar with them).

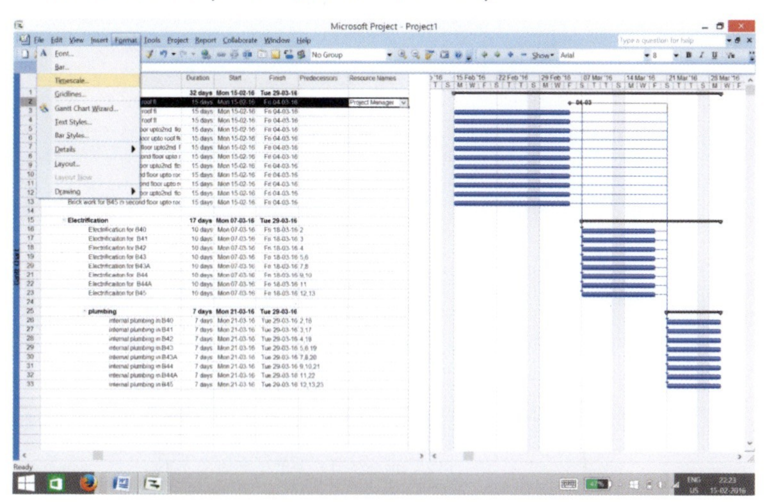

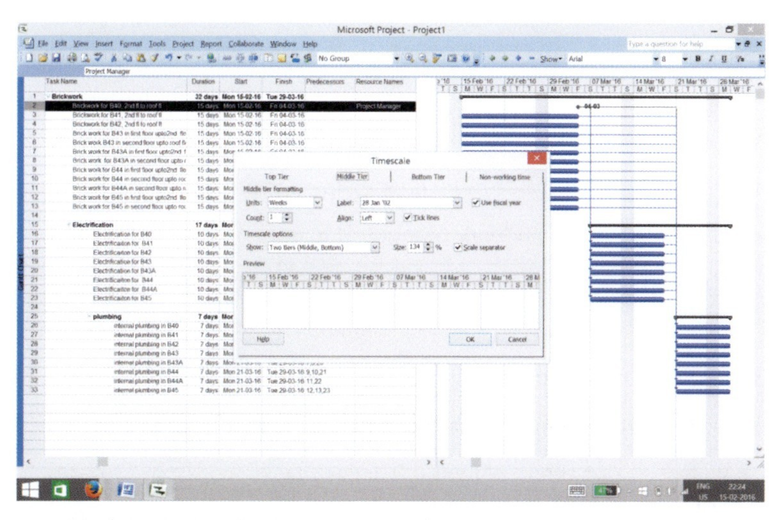

7. View the critical path

a) For the example, we are going to use the Detailed Gantt Chart to view the Critical Path, because this option also shows the Slack Time of the activities that don't belong to the critical path, therefore first we have to select the option "More Views".

b) Then we have to select the Detail Gantt to obtain the view desired (Adjust the Gantt Chart as explained before if is necessary).

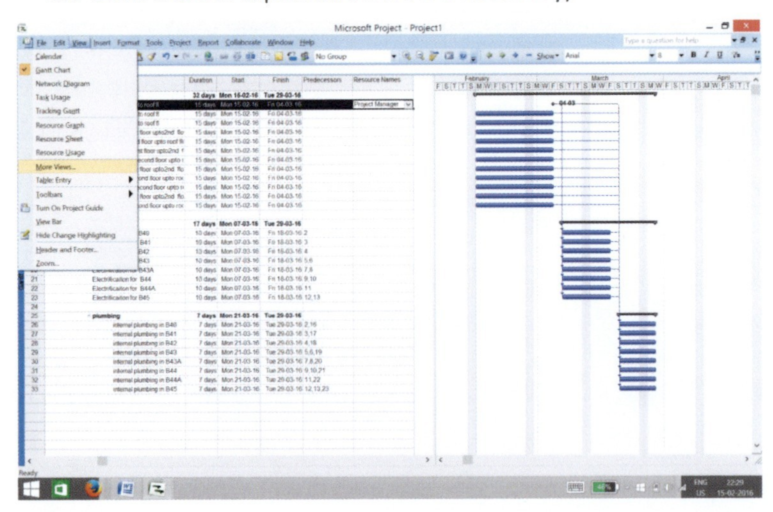

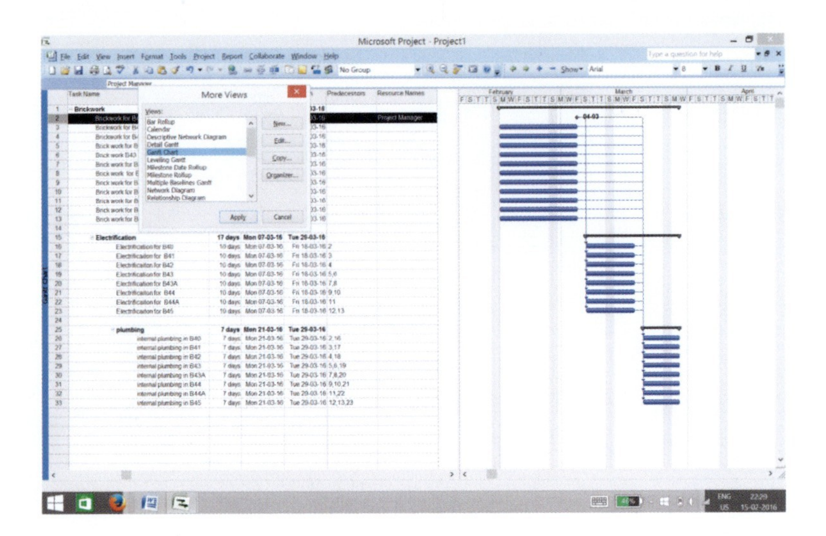

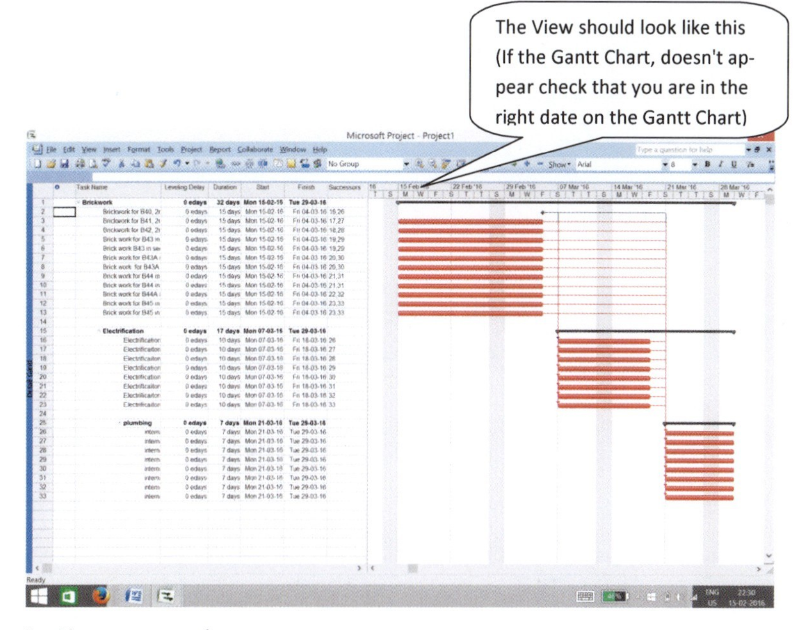

8. Show more columns

a) We can show more information, related to the tasks, in the spreadsheet, one column that might be of general interest is the cost, to do this first perform a Right Click on top of the spread sheet (Specifically In the titles of the Columns), a pop-up menu should appear showing several options, chose the one that says “Insert Column”

b) Then lookup the column named “Cost” and then press the “Ok” button.

9. As explained before, you can add and hide columns from the Spread sheet, this lets you show exactly what the people needs to see, below is a view with selected fields: Name, Cost, Duration, Resource initials and Start Date. The reader is welcome to experiment with this features and to explore more views that are offered by MS Project, such as resources usage, cost reports, etc.

10. Save the file

a) For this example we are saving the file at the end, but it is recommended that you save the file frequently while you are working to avoid losing data as a result of problems such as a Power Failure for instance.

b) You can chose between saving the file with or without Baseline (the difference was explained before in this Tutorial)

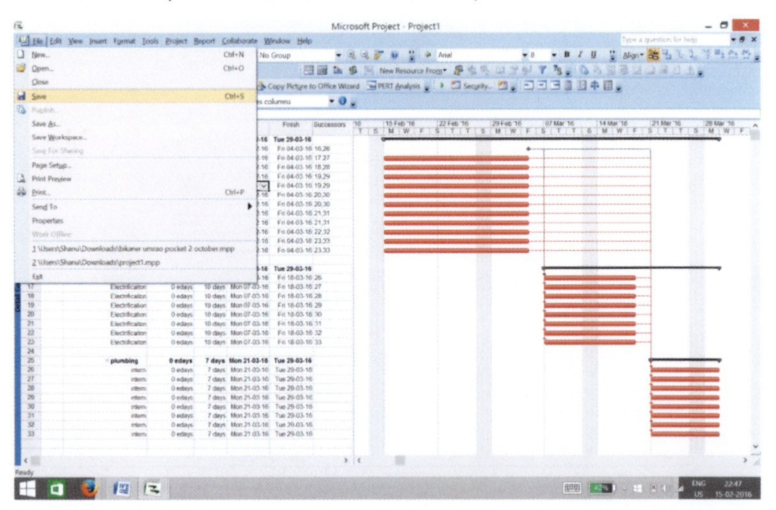

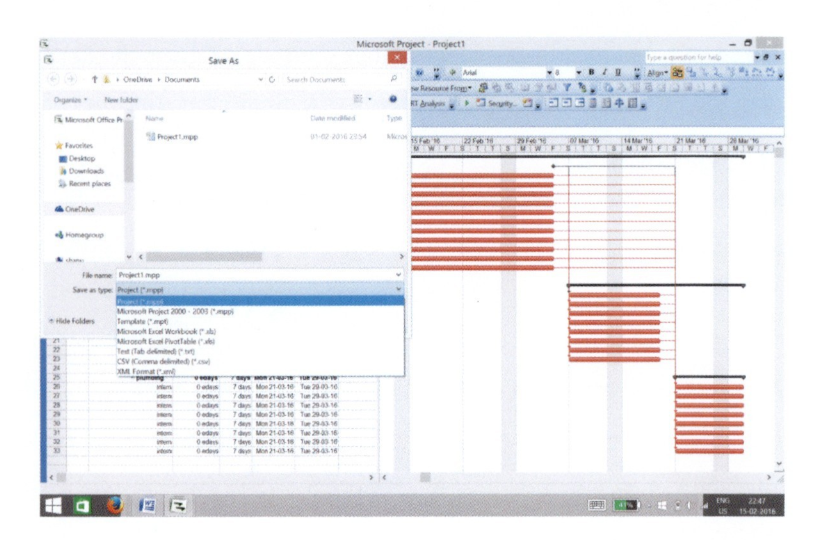

Printed in Poland
by Amazon Fulfillment
Poland Sp. z o.o., Wrocław